Airbase Vulnerability to Conventional Cruise-Missile and Ballistic-Missile AttackS

Technology, Scenarios, and U.S. Air Force Responses

John Stillion

David T. Orletsky

Prepared for the United States Air Force

Project AIR FORCE

RAND

The research reported here was sponsored by the United States Air Force under Contract F49642-96-C-0001. Further information may be obtained from the Strategic Planning Division, Directorate of Plans, Hq USAF.

Library of Congress Cataloging-in-Publication Data

Stillion, John.
Airbase vulnerability to conventional cruise-missile and ballistic-missile attacks : technology, scenarios, and U.S. Air Force responses / John Stillion and David T. Orletsky.
p. cm.
"Prepared for the U.S. Air Force by RAND's Project AIR FORCE."
"MR-1028-AF."
ISBN 0-8330-2700-X
1. Air bases—Security measures—United States. 2. Cruise missile defenses—United States. 3. Ballistic missile defenses—United States. 4. United States. Air Force—Security measures. I. Orletsky, David T., 1963- . II. United States. Air Force. III. Project AIR FORCE (U.S.).
UG634.49.S75 1999
358.4 ' 17 ' 0973—dc21 99-12338
CIP

RAND is a nonprofit institution that helps improve policy and decisionmaking through research and analysis. RAND® is a registered trademark. RAND's publications do not necessarily reflect the opinions or policies of its research sponsors.

Published 1999 by RAND
1700 Main Street, P.O. Box 2138, Santa Monica, CA 90407-2138
1333 H St., N.W., Washington, D.C. 20005-4707
RAND URL: http://www.rand.org/
To order RAND documents or to obtain additional information, contact Distribution Services: Telephone: (310) 451-7002;
Fax: (310) 451-6915; Internet: order@rand.org

PREFACE

In fiscal year 1997, under the sponsorship of the Air Force Assistant Deputy Chief of Staff for Air and Space Operations and the Air Force Director of Strategic Planning, RAND's Project AIR FORCE Strategy and Doctrine Program began a two-year effort to explore the role of air and space power in future conflicts. The primary objective of the study was to explore the prospects for developing a construct for air and space power that capitalizes on forthcoming air and space technologies and associated concepts of operation (CONOPS); that is effective against adversaries with diverse economies, cultures, political institutions, and military capabilities; and that offers an expansive concept of air and space power across the entire spectrum of conflict.

Under this broader study, the research team investigated the possibility that future adversaries might be able to mount effective missile attacks on U.S. Air Force (USAF) main operating bases in critical regions. Both emerging technologies and the proliferation of existing capabilities will give adversaries pursuing anti-access strategies a variety of new options against U.S. airbases, ports, troop concentrations, and ships at sea.

This report is not intended to assess the relative vulnerabilities of these various force elements and facilities. Rather, its purpose is to help the USAF address a potential vulnerability of its in-theater bases. The proliferation of Global Positioning System (GPS) guidance and submunition warhead technologies could make highly accurate attacks possible against USAF aircraft on parking ramps at these bases. If such attacks are feasible, the current USAF operational concept of high-tempo, parallel strikes from in-theater bases could be put in jeopardy. It is for this reason that this operational problem was deemed relevant—indeed central—to the purposes of the overall study on the future of airpower. The research documented in this report concluded that these guidance and munition technologies could, in fact, put USAF bases at serious risk. We recommend that others with expertise in land and naval operations conduct similar assessments of their vulnerabilities to these and other new technologies. The report describes the threat technologies and concept of operation in detail, then explores both short- and long-term responses to these threats.

This report should be of interest to USAF planners and operators in the Air Staff, Major Command, and Numbered Air Force Headquarters and operational units, as well as to students of air and space power in the other services and the broader defense community.

Project AIR FORCE

Project AIR FORCE, a division of RAND, is the Air Force federally funded research and development center (FFRDC) for studies and analysis. It provides the Air Force with independent analyses of policy alternatives affecting the development, employment, combat readiness, and support of current and future air and space forces. Research is performed in four programs: Aerospace Force Development; Manpower, Personnel, and Training; Resource Management; and Strategy and Doctrine.

CONTENTS

FIGURES

TABLES

SUMMARY

During the 43-day Gulf War, the U.S. Air Force (USAF) flew nearly 70,000 sorties, attacked over 28,000 targets, shot down 36 Iraqi aircraft, disrupted Iraqi command and control and transportation systems, and directly attacked the Iraqi army in Kuwait, destroying many of its vehicles and damaging its morale before the ground offensive began. All of this damage was achieved at the cost of just 14 aircraft, which were lost to ground-based air defenses; none were lost in air-to-air combat. The USAF plans to build on its success in Operation Desert Storm by deploying increasing numbers of stealthy aircraft and precision-guided munitions (PGMs), supported by a much more capable battle-management system, to fight the next war.

However, the USAF is not the only organization to have drawn lessons from Operation Desert Storm. Potential adversaries are likely to expend considerable time, energy, and resources on ensuring that the USAF does not make such a large contribution to victory at so low a cost in a future conflict. The research reported here confronts this possibility and examines ways of dealing with it.

PURPOSE AND APPROACH

Other RAND research has explored how potential adversaries could use asymmetric strategies, special operations forces, terrorists, information attacks, and weapons of mass destruction to degrade or eliminate USAF combat capability during a future conflict.[1] This report presents yet another strategy whereby a clever and competent adversary could attempt to interfere with USAF combat operations if the USAF sticks to the operational concepts that served it so well during Operation Desert Storm. It examines the following questions:

- How could potential adversaries use readily available commercial and military technology to modify conventionally armed cruise and ballistic missiles to effectively attack USAF aircraft on the ground, at theater operating bases?
- How technically advanced must an adversary be to successfully suppress USAF operations from theater operating bases?
- What options (new operational concepts, material, equipment, etc.) exist for the USAF to minimize the impact of conventional cruise- and ballistic-missile attacks on theater operating bases, both in the near term and long term?

EMERGING THREAT TECHNOLOGY

Ballistic and cruise missiles must be accurate to be militarily effective. Although many countries around the world deploy ballistic missiles similar to those used by Iraq in 1991, these weapons have limited utility against military targets. Several countries, including Iraq and Iran, have used them in combat—but only as terror weapons.[2] Technological sophistication is required for an accurate and robust (militarily suitable) cruise missile, which means that only a few nations currently possess inventories, and only the United States has used cruise missiles in combat since the end of World War II. Global Positioning System (GPS) guidance devices provide a fairly cheap and effective way of improving both ballistic- and, especially, cruise-missile guidance. This technology could be used to improve

---

[1]See David Shlapak and Alan Vick, *"Check Six begins on the ground": Responding to the Evolving Ground Threat to U.S. Air Force Bases*, Santa Monica, Calif.: RAND, MR-606-AF, 1995; Maurice Eisenstein, "The Use of Weapons of Mass Destruction by Terrorists Against Air Bases," unpublished RAND research; Brian Chow, *Air Force Operations in a Chemical and Biological Environment*, Santa Monica, Calif.: RAND, DB-189/1-AF, 1998.

[2]*Terror weapons* are weapons designed specifically to cause damage, casualties, and fear within the targeted civilian population.

the accuracy of existing ballistic missiles to about 100 meters and allow almost all nations to obtain the accurate cruise missiles that, until now, have been reserved for technologically advanced societies.

However, improved missile accuracy is not enough to make ballistic and cruise missiles both militarily effective and affordable weapons against parked aircraft. Submunitions are far more efficient against soft targets susceptible to blast or fragmentation damage than are unitary warheads of the same weight. The lethal area of a cruise missile with a 75-pound payload against aircraft in the open is about three times greater when using a submunition warhead than when using a unitary warhead. This advantage increases with increasing payload. An 1,100-pound M-9 ballistic-missile warhead covers almost eight times the area when using a submunition warhead than when using a unitary warhead.[3] The *combination* of increased accuracy from GPS guidance and increased warhead efficiency is what decreases the number of missiles required to attack USAF airbases from hundreds to dozens.

A potential asymmetric strategy considered in this report is the use of small, slow cruise missiles to "slip under" the current USAF radar umbrella. The term "cruise missile" simply refers to an unmanned aircraft designed to fly a one-way attack mission. The cruise missiles considered here are significantly different from the high-performance fighter-size targets USAF air defense systems were designed to counter during the Cold War. Cruising at about 70 knots, these small aircraft would be difficult for current USAF air defense systems to detect. Surveillance and tracking radars designed during the Cold War (e.g., those on F-15s, F-16s, and Airborne Warning and Control System [AWACS]) took advantage of the high speed of Soviet combat aircraft to simplify the task of sorting attacking aircraft from ground-vehicle clutter merely by ignoring potential targets moving slower than about 80 knots. Some of the systems have the capability to detect and track slower targets, but only in narrow sectors and for short periods of time before the number of potential targets exceeds the system's data-processing and display capabilities.

---

[3]These calculations assume a 20-foot lethal radius for a 1-pound submunition and that 75 percent of warhead weight is devoted to submunitions, with the remainder devoted to a frame and dispensing mechanism.

Surface radars are less affected by ground clutter than are airborne radars but suffer from limited line of sight against low-flying targets. Patriot and AEGIS[4] could acquire and track a slow-moving cruise missile, but only above the radar horizon—less than 20 miles for a cruise missile flying at 100 to 130 feet. Unless the United States deploys huge numbers of ground-based radars to a future theater conflict, most cruise missiles will go undetected by current U.S. air defense systems.

OPERATIONAL IMPACT

We posit a simple illustrative scenario to explore the impact GPS-guided cruise and ballistic missiles equipped with submunition payloads might have on current USAF theater air operations. In our scenario, Iran uses an Iraqi succession crisis turned civil war as an opportunity to invade southern Iraq. The United States responds in a variety of ways, including deploying USAF combat aircraft to the following bases on the Arabian peninsula: Dhahran, Doha, Riyadh Military, and Al Kharj.

These bases have a total of 14 potential parking areas ranging in size from 600 × 300 feet to 9,000 × 900 feet. The total area of the parking ramps at these bases is over 44 million square feet—the equivalent of almost 1,000 football fields. These bases can accommodate a huge number of combat aircraft and an intense aerial-port operation. However, the number of GPS-guided, submunition warhead cruise missiles and ballistic missiles required to attack this huge area is surprisingly small, assuming a 20-foot lethal radius for the 1-pound submunitions employed and standard USAF aircraft-parking procedures. A 0.9 Pk (probability of kill) against all aircraft on the parking ramps of these four bases could be achieved with 30 GPS-guided M-9 and 30 M-18 ballistic missiles, and 38 small GPS-guided cruise missiles, at an estimated cost of about $101 million.

---

[4]AEGIS is a totally integrated shipboard weapon system that combines computers, radars, and missiles to provide a defense umbrella for surface shipping. The system is capable of automatically detecting, tracking, and destroying airborne, seaborne, and land-launched weapons. Joint Chiefs of Staff, *Department of Defense Dictionary of Military Terms*, Washington, D.C.: Joint-Pub 1-02, March 23, 1994, pp. 6–7.

Attacking the tent cities at all four bases and a Patriot or theater high-altitude air defense (THAAD) radar at each requires an additional 40 ballistic missiles and 8 cruise missiles, raising the total cost to about $163 million—about the cost of four Russian Su-27 export-version fighters. The effect on USAF sortie generation of destroying a large number of aircraft, living quarters, most personal equipment, and some work centers while creating widespread foreign-object damage would be devastating.

POSSIBLE USAF RESPONSES

To reduce the vulnerability of deployed forces, the USAF could take a variety of actions over the next few years. These actions fall into three basic categories: passive defenses, active defenses, and dispersal.

Passive defenses include constructing hardened aircraft shelters and living facilities at likely deployment bases; acquiring deployable shelters, for both aircraft and personnel, capable of withstanding submunition impact; and constructing additional parking-ramp space to allow increased dispersal. These measures would complicate an adversary's targeting problem and increase the number of weapons required to achieve a given level of damage.

All of these measures could be effective against GPS-guided cruise- and ballistic-missile attacks, but have potentially serious drawbacks. Hardened shelters and additional parking ramps are expensive, time-consuming construction projects that require the USAF to correctly anticipate—years in advance—where it will fight the next war. Deployable shelters allow more-flexible operations, but significantly increase the wing's airlift requirements.

Short-term *active defenses* against the small, slow cruise-missile threat could include relatively low-tech, simple measures such as putting machine-gun teams with night-vision goggles in towers surrounding USAF operating bases or deploying radar-guided guns.

Another relatively short-term alternative available to the USAF is to *disperse* its operations to a large number of highway landing strips to complicate an adversary's targeting problem. This option has the potential to defeat the missile threat but, again, carries significant potential costs, especially for sortie-generation activities. Sortie rate would not necessarily be reduced by dispersed operations, given the economies-of-scale considerations in maintenance and force protection, but would require more personnel to achieve the same number of sorties (all other things being equal—range to target, availability of munitions, etc.). In addition, an adversary with access to effective human intelligence (HUMINT) or satellite capability could locate and attack USAF units at these dispersed locations.

The Expeditionary Air Force (EAF) concept also must be considered when formulating basing operations and vulnerability to missile attack. This concept emphasizes the ability to rapidly deploy anywhere in the world, which raises two issues for defense planning: First, little support will exist to build additional infrastructure (shelters, additional ramp space, etc.) at potential deployment bases that could reduce the impact of airbase attack. Second, since the EAF must travel light to deploy a warfighting package quickly anywhere in the world, little flexibility will exist to transport items that would provide protection or facilitate recovery from such attacks.

POTENTIAL LONG-TERM SOLUTIONS

If the USAF could conduct its theater air campaigns from a few secure bases with assured access,[5] many of the issues discussed in this report could be avoided. Hardened facilities and advanced missile defenses could be constructed prior to the start of hostilities. Although a detailed cost analysis was beyond the scope of this work, building a robust passive and active defense system at only a few selected bases should limit the total expense.

To achieve the goal of operating anywhere in the world from a few secure, hardened, fixed bases with guaranteed access, the USAF would need to develop operational concepts for longer ranges. Crew-fatigue considerations limit aircraft with 500-knot cruise speeds to ranges of about 2,000 nautical miles (nmi) for sustained operations. An aircraft with a 1,000-knot cruise speed could cover virtually the entire inhabited land surface of the Earth, except for the

---

[5]*Assured access* means basing in the United States or on the territory of very close allies, such as the United Kingdom, who typically support U.S. actions.

southern tips of Africa and South America, by operating from four secure hardened bases: Guam, near Anchorage, Miami, and London.

A total inventory of approximately 80 to 105 Mach 2 bombers with the following specifications could deliver enough PGMs (about 560 tons per day) to replicate the USAF Desert Storm effort:

- a weight of 290,000 to 350,000 pounds
- an unrefueled range of 3,250 nmi
- a payload of 15,000 to 20,000 pounds.

This force could attack targets almost anywhere in the world while operating from well-protected, permanent bases in the United States and the United Kingdom.

ACKNOWLEDGMENTS

The authors thank their RAND colleague and project leader, Alan Vick, for suggesting the topic and providing guidance and assistance throughout the effort. We are grateful to LtCol Julie Neumann, who served as action officer for the project. We also thank our RAND colleagues Irving Lachow and John Pinder for their assistance. Tim Bonds and David Shlapak reviewed a draft of this document. Many of their suggested changes have been incorporated here, and have improved the substance and style of this document. Finally, we extend our deepest and most grateful appreciation to our editor, Marian Branch. Once again, she has worked the magic we have become so accustomed to, significantly improving the quality of this manuscript.

ABBREVIATIONS AND ACRONYMS

ABM	anti–ballistic missile (system)
AEF	Air Expeditionary Force
AFB	Air Force Base
AFH	Air Force Handbook
APOD	aerial port of debarkation
AWACS	Airborne Warning and Control System
BM	ballistic missile
C/A	coarse acquisition (navigation signal—GPS)
CEP	circular error probable
CM	cruise missile
CONOPS	Concept of operations
CRAF	Civilian Reserve Airlift Fleet
CSS	Chinese surface-to-surface (missile)
DAFIF	Digital Aeronautical Flight Information File
DE	directed energy
DGPS	differential Global Positioning System
DOPS	Defense Intelligence Agency Outline Plotting System
EAF	Expeditionary Air Force
FOB	Forward Operating Base
FOD	foreign-object damage

FT	flight time
GPS	Global Positioning System
GT	ground time
GWAPS	*Gulf War Air Power Survey*
HE	high explosive
HUMINT	human intelligence
INS	inertial navigation system
IOC	initial operational capability
IR	infrared
IRSTS	Infra-Red Search and Track System
ISR	intelligence, surveillance, reconnaissance
JDAM	Joint Direct Attack Munition
JSTARS	Joint Surveillance, Targeting, and Reconnaissance System
L/D	lift divided by drag
MLRS	multiple-launch rocket system
MOB	Main Operating Base
MT	maintenance time
MTW	major theater warfare
NATO	North Atlantic Treaty Organization
PGM	precision-guided munition
Pk	probability of kill
PRC	People's Republic of China
RAF	Royal Air Force (United Kingdom)
rms	root mean square
SA	Selected Availability
SAM	surface-to-air missile
SDPR	System Design and Performance Requirements
SEAD	suppression of enemy air defense
SHORAD	short-range air defense

SOF	special operations force
SPS	Standard Positioning Service (GPS)
SR	sortie rate
SRBM	short-range ballistic missile
STTO	start, taxi, take off
SWA	Southwest Asia
TAT	turnaround time
TERCOM	TERrain COntour Matching
THAAD	theater high-altitude air defense
TLE	target-location error
UAE	United Arab Emirates
UAV	unmanned aerial vehicle
USAF	United States Air Force

Chapter One

INTRODUCTION

The tremendous success of U.S. Air Force (USAF) operations against the Iraqi military during the 1991 Gulf War was a stunning achievement. Videotaped images of laser-guided bombs being dropped from USAF stealth aircraft with impunity onto targets deep inside Iraq contain the seeds of both the popular perception of how the war was fought and the "lessons learned" the USAF took away from the war. During the course of the 43-day war, the USAF flew 69,406 sorties, attacked 28,295 targets, shot down 36 Iraqi aircraft, disrupted Iraqi command and control and transportation systems, and directly attacked the Iraqi army in Kuwait, destroying many of its vehicles and severely damaging its morale before the ground offensive began. All of this was achieved at the cost of just 14 aircraft, which were lost to ground-based air defenses, not in air-to-air combat.[1]

The USAF's success in Operation Desert Storm changed the way military commanders and the general public view airpower in particular and modern warfare in general. The public expectation is now that the USAF could attack almost any target at any time and effectively destroy it, with little risk to U.S. military personnel or to noncombatants on the ground. As well, the war left the impression that, while certain aspects of USAF operations such as bomb-damage assessment could be refined to improve the efficiency of operations, the basic concept of deploying large numbers of combat aircraft and support personnel to vulnerable bases within a few hundred nautical

[1]Eliot A. Cohen, ed., *Gulf War Air Power Survey* [GWAPS], *Vol. V: A Statistical Compendium and Chronology*, Washington, D.C.: U.S. Government Printing Office, 1993, pp. 316, 418, 651–654.

miles of enemy territory should continue to be the blueprint for future USAF theater air campaigns. However, these operations were conducted against an adversary who lacked the means, will, skill, and vision to attempt to attack and disrupt USAF aircraft and operations where they are most vulnerable—at their fixed operating bases. Next time out, the USAF may not be so lucky.

In fact, the air campaign seems to have achieved so much at so small a cost that potential adversaries are likely to expend substantial time, energy, and resources to ensure that the USAF does not make such a large contribution to victory at so low a cost in a future conflict.

PURPOSE

Other RAND research has explored how potential adversaries could use asymmetric strategies, special operations forces, terrorists, information attacks, and weapons of mass destruction to degrade or eliminate USAF combat capability during a future conflict.[2] This report presents yet another way in which a clever and competent adversary could attempt to interfere with USAF combat operations if the USAF sticks to the operational concepts that served it so well during Operation Desert Storm: modifying cruise and ballistic missiles to attack USAF aircraft on the ground.

The report examines the following questions:

- How could potential adversaries use readily available commercial and military technology to modify conventionally armed cruise and ballistic missiles to effectively attack USAF aircraft on the ground, at theater operating bases?
- How technically advanced must an adversary be to successfully suppress USAF operations from theater operating bases?

---

[2]See David Shlapak and Alan Vick, *"Check Six begins on the ground": Responding to the Evolving Ground Threat to U.S. Air Force Bases*, Santa Monica, Calif.: RAND, MR-606-AF, 1995; Maurice Eisenstein, "The Use of Weapons of Mass Destruction by Terrorists Against Air Bases," unpublished RAND research; Brian Chow, *Air Force Operations in a Chemical and Biological Environment*, Santa Monica, Calif.: RAND, DB-189/1-AF, 1998.

- What options (new operational concepts, material, equipment, etc.) exist for the USAF to minimize the impact of conventional cruise- and ballistic-missile attacks on theater operating bases, both in the near term and long term?

The answers presented here are speculative. The analysis is based on potentially available technologies and an understanding of USAF operations and is intended to expose a potential vulnerability that could be exploited by future adversaries. Since many of the weapons discussed in this document would appear to have shorter and less costly design and production cycles than USAF aircraft have, these threats should be considered prior to acquiring aircraft. Throughout this document, we use open source material and simple calculations to develop a potentially serious threat to USAF aircraft that many conceivable adversaries could institute.

The Expeditionary Air Force (EAF) concept also must be considered when formulating basing operations and vulnerability to missile attack. This concept emphasizes the ability to rapidly deploy anywhere in the world, which raises two issues for defense planning: First, little support will exist to build additional infrastructure (shelters, additional ramp space, etc.) that could reduce the impact of airbase attack at potential deployment bases. Second, since the EAF must travel light to deploy a warfighting package quickly anywhere in the world, little flexibility will exist to transport items that would provide protection or facilitate recovery from such attacks.

ORGANIZATION

Chapter Two describes how existing, readily available technologies could enable a future regional competitor to build a cheap, accurate, and effective cruise- and ballistic-missile force to attack USAF aircraft where and when they are most vulnerable—on the ground at theater operating bases. Chapter Three presents a scenario to illustrate how an adversary might employ conventional cruise and ballistic missiles to attack USAF theater operating bases, and the potential impact these attacks might have on USAF operations. Chapter Four describes several short- to medium-term actions the USAF could take to counter the conventional cruise- and ballistic-missile threat. Examples of these actions include increased efforts, both active and passive, to defend USAF theater operating bases from missile attack. Chapter Five describes a new short-term operational concept that could allow the USAF to operate effectively in the face of our postulated threat. In addition, it outlines a longer-term strategy the USAF could adopt: By dramatically increasing the speed and/or range of its ground attack aircraft, the USAF could minimize its vulnerability to conventional missile attacks. This improvement would allow the USAF to conduct a sustained air campaign against targets almost anywhere in the world from a handful of well-defended bases located on U.S. soil or controlled by the United States' most reliable allies. Chapter Six presents conclusions. Appendix A presents additional data and assumptions we used to model adversary missile requirements. Appendix B presents the assumptions and equations that form the basis of our sortie-rate-generation model, which informs the analysis of Appendix C. Appendix C gives some of the technical details of one possible long-range-attack-aircraft–weapons combination.

Chapter Two

EMERGING THREAT TECHNOLOGIES

"The large ground organization of a modern air force is its Achilles' heel."

Basil Henry Liddell Hart, *Thoughts on War,* 1944

THE EAGLES IN THEIR NESTS

The notion that military aircraft are most vulnerable when they are on the ground is nothing new. Throughout World War II, both sides used surprise air attacks, resistance units, special operations forces, and even nighttime naval bombardment to attack parked aircraft. In some instances, such as the Japanese air attack on Pearl Harbor; the initial German air assaults on the Polish, Dutch, Belgian, French, and Russian air forces; and the British long-range desert group operations against Axis airfields in North Africa,[1] spectacular results were achieved.

The Iraqis failed to launch a concerted campaign to attack USAF aircraft at their bases during the 1991 Gulf War. The photograph in Figure 2.1 shows just how wide open [literally] and vulnerable many

[1]For a detailed and fascinating account of these little-known operations, see Alan Vick, *Snakes in the Eagle's Nest: A History of Ground Attacks on Air Bases*, Santa Monica, Calif.: RAND, MR-553-AF, 1995, Chapter Four. Obviously, the authors have borrowed the excellent imagery evoked by the title of Alan Vick's work in titling this section of the chapter.

parked U.S. aircraft were during that conflict. The Iraqi failure, most probably, is the result of multiple factors:

- Coalition air defenses were so powerful and efficient that the Iraqi Air Force lacked the size, training, and equipment to launch anything more than a one-time suicidal raid on coalition air bases in Saudi Arabia.
- Iraqi ballistic missiles were not accurate enough to have more than a trivial probability of hitting the typical airbase aircraft parking ramp. Even if lucky enough to land on the ramp, the unitary warheads on the Iraqi missiles were not capable of damaging more than a few parked fighter aircraft.
- Harder to explain is a third factor: Iraq was unable to organize or hire special operations forces or terrorists to attack USAF aircraft as the North Vietnamese did two decades earlier.

USAF photo, courtesy of DoD Still Media Records Center

Figure 2.1—Parking Ramps at Shaikh Isa, Bahrain, Early 1991

Regardless of the reasons for the lack of Iraqi airbase attacks, there is little doubt that an adversary who could successfully mount such attacks would present a difficult challenge. The Iraqis, and other regional military powers who believe they may someday confront U.S. airpower in a crisis or conflict, are almost certainly searching for inexpensive but effective ways to enhance their capability to attack USAF assets where they are most vulnerable—on the ground.

In the rest of this chapter, we explain how a powerful regional competitor could combine readily available ballistic-missile, unmanned aerial vehicle (UAV), Global Positioning System (GPS), and submunition warhead technologies to produce a variety of accurate, effective, and inexpensive missiles capable of devastating attacks on USAF aircraft parked on open ramps. First, we discuss the cruise- and ballistic-missile threats and enhancements that could be incorporated to improve accuracy and lethality. We then discuss the potential effectiveness of these weapons against USAF aircraft.

POTENTIAL BALLISTIC- AND CRUISE-MISSILE THREATS

Ballistic and cruise missiles were first used in combat by the Germans during the last year of World War II. Since then, ballistic missiles have been produced in many countries and purchased by many more.[2] Ballistic missiles were used extensively by both sides in the Iran-Iraq War during the 1980s, and most recently by Iraq in the 1991 Gulf War against targets in Israel and Saudi Arabia. For the most part, the Scud missiles used in these recent conflicts differed little from their German V-2 ancestors: They had about the same range, unitary high-explosive payload, and accuracy.

This lack of refinement meant that they could attack the same type of large-area targets—mostly cities—that the German missiles had been used against. In the 1991 Gulf War, Scud attacks against cities could have had important political ramifications (e.g., bringing Israel into

---

[2]The following countries have ballistic missiles that can deliver at least a 500-kilogram (kg) payload to a target at a 500-kilometer (km) range: Brazil, People's Republic of China (PRC), Egypt, India, Iran, Israel, Japan, North Korea, Libya, Pakistan, Saudi Arabia, South Africa, Spain, Syria, Taiwan, and Zaire. See *The Nonproliferation Review*, Spring–Summer 1995, Vol. 2, No. 3, pp. 203–206. France, Russia, the United Kingdom, and the United States also possess missiles with at least this capability.

the conflict) and caused the coalition to expend significant resources in deploying additional Patriot missile batteries and hunting mobile Scud launchers in the desert wastes of Iraq. However, these indirect effects were far too small to alter the overall military outcome. To be militarily effective, ballistic missiles must be far more accurate and have much more efficient warheads than were the Scuds employed by Iraq in 1991.

In contrast to the relatively wide proliferation of ballistic missiles since the end of World War II, land-attack cruise missiles have been deployed and used in combat by only one nation—the United States. The primary reason for this exclusivity is that only the United States had the technical capability and financial resources to solve the problems associated with accurate cruise-missile navigation.

IMPROVING ACCURACY WITH GPS

The accuracy of a missile is typically expressed as a *circular error probable* (CEP), which is defined as the radius of a circle within which half of the missiles land for a given aimpoint. This parameter works well for calculating the probability of kill or the number of weapons required to destroy a target. But a different description of missile error is needed to assess the impact of enhanced guidance systems, because several error sources affect the accuracy of missiles. And because the total guidance error is the square root of the sum of the squares of the individual errors, total system inaccuracy is determined to a great extent by the single largest error source. The three major categories of guidance error are errors in launch position accuracy, en route errors, and target-location errors.

The Global Positioning System is not a magic technology that solves all accuracy problems. Rather, it has the potential to enhance flexibility of operations and improve accuracy by reducing target-location error (TLE) for both ballistic and cruise missiles. Target coordinates obtained from GPS-equipped intelligence sources (human agents, stand-off reconnaissance aircraft, satellites, ships, etc.) will probably be more accurate because the acquiring platform (or people) knows (know) more precisely where it (or they) is (are). Launch-position errors, on the contrary, may not be significantly reduced, since countries typically use pre-identified and surveyed launch sites. The advantage of using GPS is that it enables more-flexible launch operations to be conducted. Missiles can be accurately launched from anywhere, including offshore platforms, making the launcher more difficult to engage. Another advantage that GPS provides to both types of missiles is a consistent coordinate system, which eliminates transformation errors that can accrue as coordinates are switched from one grid to another.

Because the two missiles have different characteristics, primarily time of flight, the addition of GPS guidance[3] enhances the accuracy of ballistic and cruise missiles in different ways, discussed in the following two subsections.

Ballistic Missiles

Errors in the boost and reentry phases of Third World ballistic missiles arise primarily from factors such as variance between commands and actual engine cutoff and aerodynamic forces during reentry. These errors will be virtually unaffected by GPS assistance. The guidance system can send precise engine-cutoff information, but if the engines are not sufficiently advanced to shut down at precisely the right moment, the missile accuracy will not be significantly improved.

Compensating for errors arising from aerodynamic forces during reentry requires that reentry vehicles be equipped with control surfaces or thrusters to correct for wind drift; most existing short-range ballistic missiles (SRBMs) lack these systems. As a result, adding GPS guidance to a Scud missile improves accuracy by only about 20 percent (bringing the CEP to about 600 meters). In contrast, an advanced SRBM, such as the Chinese M-9, which has warhead-control thrusters, could achieve CEPs in the neighborhood of 150 to 200

---

[3]*GPS guidance* as used in this document means *integrated GPS–inertial navigation system (INS) guidance*. Inertial navigation systems use gyroscopes to compute position and, therefore, the accuracy of the computed position is inversely related to the time of flight. An integrated GPS–INS system updates the inertially computed position with GPS information. Therefore, the positional accuracy is independent of time of flight. The integrated GPS–INS system prevents the defender from spoofing the system by jamming the GPS signal in the neighborhood of the target.

meters, depending on the accuracy of the GPS signal used (SPS, C/A, or DGPS).[4]

Cruise Missiles

Unlike ballistic missiles, whose flight times are measured in minutes, long-range cruise missiles can take several hours to reach their targets. Long flight times within the atmosphere, and the attendant large effect of unpredictable winds on the missile's course, have been the major guidance problem facing cruise-missile designers since the 1940s. Therefore, GPS guidance has tremendous potential for reducing cruise-missile en route navigation errors.

Even the best inertial navigation systems (INS) are insufficiently precise to guarantee pinpoint accuracy over extended periods. Imagine a cruise missile equipped with an INS featuring a drift rate of 0.1 mile per hour (mph)—roughly characteristic of high-quality current-generation systems. Flying at 350 knots, a typical speed, the missile would require about 2.85 hours to fly a 1,000-nautical-mile (nmi) mission. Over that time, the INS would introduce an error of almost three-tenths of a nautical mile, or about 1,500 feet (463 meters) into the missile's path. Left unassisted, the result would be a weapon with insufficient accuracy for conventional attacks against anything but large-area targets. Midcourse updates to the INS can eliminate the accrued drift errors and greatly improve guidance-system performance under such circumstances.

The United States solved the in-flight-update problem by equipping its land-attack cruise missiles with TERain COntour Matching (TERCOM) guidance systems developed during the 1970s. To determine exactly where the cruise missile is in relation to its planned course, these systems compare radar altimeter readings of the terrain a missile is passing over with stored digital maps created from

---

[4]Our colleagues Gerald Frost and Irving Lachow provide a useful discussion of these issues in a series of documents on the subject. Two directly related to this discussion are *Satellite Navigation-Aiding for Ballistic and Cruise Missiles*, Santa Monica, Calif.: RAND, RP-543, 1996a; and *GPS-Aided Guidance for Ballistic Missile Applications: An Assessment*, Santa Monica, Calif.: RAND, RP 474-1, 1996b. Much of the above discussion was drawn from these documents.

reconnaissance satellite images. The guidance system can then generate corrections as necessary. This solution is complex, expensive, and requires access to still more-complex and more-expensive systems, such as reconnaissance satellites, to make the missile work. Only relatively rich nations could afford to employ such a system to solve the cruise missile–guidance problem.

Today, however, fairly simple, cheap, and widely available technology can be obtained for reducing cruise-missile en route navigation errors and accurately guiding cruise missiles to distant targets—the GPS. GPS-guided cruise missiles have the potential for achieving pinpoint accuracy. Unlike ballistic missiles, which are often unguided after engine cutoff, cruise missiles can make continuous course corrections all the way to the target. As a result, GPS-guided cruise missiles should be able to achieve accuracies near that of the GPS signal.

IMPROVING LETHALITY

GPS guidance technology offers a cheap means to improve ballistic-missile accuracy and to solve the cruise missile–guidance problem. Nevertheless, it is not enough to enable an adversary to effectively attack USAF aircraft parked in the open at theater bases. Using unitary warheads, even on accurate missiles, to attack large aircraft-parking ramps would require many hundreds or even thousands of missiles.

Submunition warhead technology provides a solution to this problem. The lethal radius of a high-explosive (HE) warhead is proportional to the cube root of the explosive weight. Eight times more explosive is required to double the lethal radius. Therefore, when attacking large, soft-skinned targets, many little explosions spread around the area are preferable to one big detonation of equivalent weight, making submunitions much more efficient than unitary warheads of the same weight against soft targets susceptible to blast or fragmentation damage (e.g., trucks, aircraft, personnel in the open, tents, radar sites). Figure 2.2 illustrates the enhanced effectiveness of submunition warheads relative to that of unitary warheads of

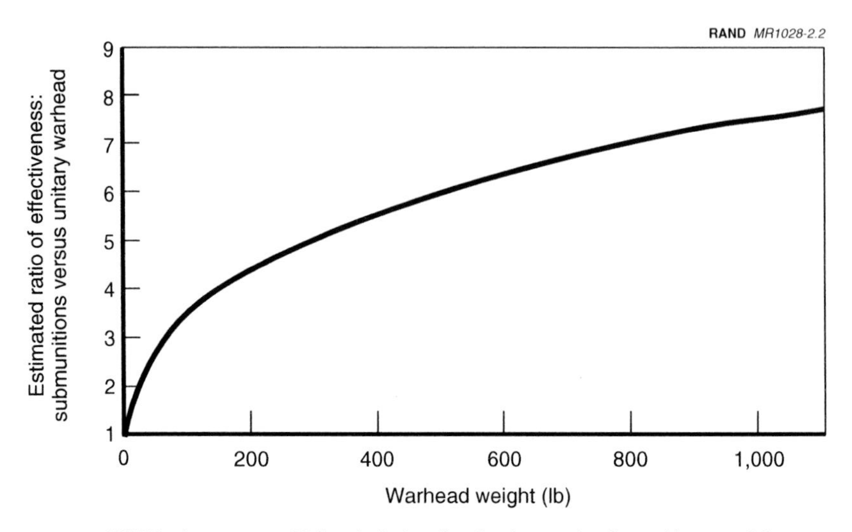

NOTE: Assumes a 20-foot lethal radius for 1-pound submunitions and that 75 percent of warhead is devoted to submunitions and the remainder to a frame and dispensing mechanism.

Figure 2.2—Lethal Area of 1-Pound Submunition Versus Unitary Warheads of the Same Weight

equivalent weight.[5] It also shows that the larger the warhead, the greater the advantage of using a submunition warhead.[6]

---

[5]For example, suppose a 1-pound submunition has a lethal radius of about 20 feet against parked aircraft. The area swept by this submunition is $3.14159 \times 20^2 = 1{,}256.6$ square feet. For comparison, using the cube rule, a unitary warhead of 1,100 pounds would have a lethal radius against parked aircraft of about $20 \times 1{,}100^{1/3} = 206$ feet—giving a lethal area of approximately 125,000 square feet. If we assume an 1,100-pound warhead could accommodate and dispense approximately 825 1-pound submunitions, its lethal area would be $825 \times 1{,}256.6 = 1{,}036{,}695$ square feet—almost eight times the lethal area of a unitary warhead of the same weight. These calculations assume 1-pound submunitions with approximately 25 percent of submunition warhead weight devoted to a dispensing mechanism and about 75 percent to bomblets. These weight fractions are typical for current U.S. cluster munitions, including bombs and multiple-launch rocket system (MLRS) rockets.

[6]Some final points about submunition technology. First, it is not a recent development. The United States has deployed bombs employing submunitions since the 1960s. Submunition warheads are well within the technological reach of many nations, and many have produced air-deliverable weapons. According to Duncan S. Lennox and Arthur Rees, eds., *Jane's Air Launched Weapons*, Surrey, England: Jane's

But how effective could these missiles be? Figure 2.3 shows the lethal areas of the unitary and postulated submunition warheads of both of the potential cruise missiles discussed in the next section and, for comparison, a Chinese M-9 or M-18 ballistic missile. The M-9 and M-18 are closely related systems. The M-18 is an M-9 with an extra booster stage for extended range. Both are fairly recent designs and incorporate a detachable warhead with control jets. The control jets enable the warhead to make steering corrections from separation to impact.[7]

To show just how much damage accurate GPS-guided missiles armed with submunition warheads could cause to aircraft in the open, 96 F-15–size fighters at typical parking-ramp spacing have been superimposed over the lethal areas of the various warheads. The small filled circles represent the lethal areas of each individual 1-pound bomblet from the missile warheads. We assume that the submunitions are dispensed with roughly 20 feet between the outer limits of each bomblet's lethal radius. Therefore, about half the wingspan of a typical fighter aircraft separates any two of the bomblets, ensuring that no parked jet can escape damage by finding itself "squeezed in" between the lethal radii of two adjacent bomblets. This fact further enhances the effectiveness of the submunition warheads.

Figure 2.3 makes it clear that less than a dozen cruise missiles like those discussed in the next section could severely damage or destroy almost an entire fighter wing parked in the open. Only one GPS-guided ballistic missile with conventional submunitions, like either the M-9 or M-18, would be required to do the same damage.

Information Group, Issue 12, 1990, the following nations have at least one air-deliverable cluster munition in production and service: Chile, China, France, Germany, Iraq, Israel, Italy, Poland, South Africa, Spain, UK, USA, USSR (now Russia), and Yugoslavia. Second, the U.S. Army MLRS uses unguided ballistic rockets to deliver large numbers of submunitions over a wide area. This technology will be about the same vintage late in the next decade as ballistic-missile technology was in the mid-1980s, when Iran and Iraq began to first purchase, and then manufacture, ballistic missiles for use in the sustained "war of the cities." It seems that any nation powerful enough to contemplate a conventional military campaign against U.S. interests or allies probably will have access to submunition-warhead technology within a decade, if it does not already.

[7]Duncan Lennox, ed., *Jane's Strategic Weapon Systems*, Coulsdon, Surrey, England: Jane's Information Group, Issue 24, May 1997.

Many nations manufacture and sell ballistic missiles on the world market. The technology and expertise to produce these missiles are also for sale. Before the Cold War ended, the Soviet Union was the primary exporter of ballistic missiles and manufacturing technology to the Third World. In recent years, China has become much more active in this area.

It would be fairly simple for any nation seeking to acquire ballistic missiles with characteristics similar to the Chinese-designed M-9 and M-18 to do so. As Figure 2.3 illustrates, modifying these missiles to carry submunition warheads would enable them to attack truly enormous areas. These missiles have much larger payloads than do the proposed cruise missiles described below. Their 500-kg payload

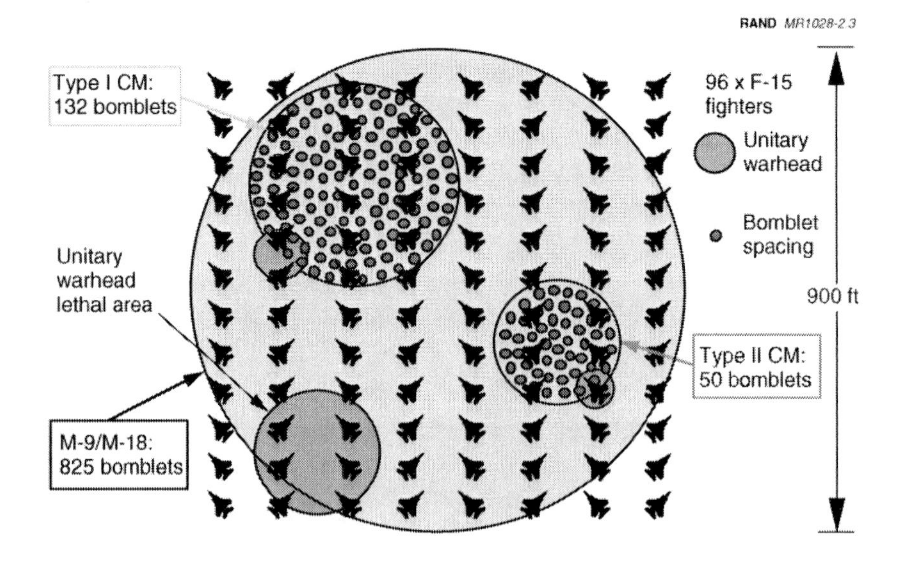

NOTE: Aircraft spacing reflects guidelines set out in U.S. Air Force, *Facility Requirements,* Air Force Handbook AFH 32-1084, September 1, 1996, Table 2.6, for F-15 aircraft parked at a 45-degree angle.

Figure 2.3—Comparison of Warhead Lethal Radii for a Typical Ballistic Missile and the Two Postulated Cruise-Missile Candidate UAVs

capacity could be used to dispense more than 800 bomblets—damaging aircraft over an area of more than 2 million square feet.[8]

THE LOW, SLOW KILLER—A CHEAP AND EFFECTIVE ANTI-AIRFIELD CRUISE MISSILE

When most people hear the term "cruise missile," they think of the high-tech, expensive Tomahawk missiles fired against Iraq by the United States during Desert Storm (and several times since). However, a cruise missile is simply an unmanned aircraft designed to fly a one-way attack mission. Nothing says it has to be jet-powered or rocket-powered, fly faster than 250 knots, or use expensive, complex guidance schemes. Relative to the Tomahawk, it can just as easily be a small, cheap, piston-engined propeller aircraft using commercially available GPS and computer technology for guidance and control. The following is a description of what we feel could be remarkably simple, affordable, effective, and survivable cruise missiles.

As discussed earlier in this chapter, GPS has dramatically simplified the problem of cruise-missile guidance. Adding a guidance package—consisting of a GPS receiver linked, perhaps, to a laser or radar altimeter—to an existing reconnaissance UAV airframe would give that craft sufficient accuracy to function as a weapon. Replacing its payload of sensors and data links with submunitions and extra fuel could result in a lethal and inexpensive (relative to the Tomahawk) cruise missile.[9]

We chose two existing reconnaissance UAVs as potential candidates for conversion to cruise missiles: the Lear R4E Skyeye, manufactured in Santa Monica, California, and the SDPR VBL-2000, manufactured in Belgrade, Yugoslavia. Table 2.1 lists the range, cruising speed,

---

[8]Available technology and techniques should suffice to enable ballistic-missile warheads to achieve the kind of well-behaved (i.e., uniform, with a high degree of certainty) submunition-distribution patterns shown in Figure 2.3.

[9]A bewildering array of small UAVs suitable for such a conversion is available on the world market. There are no export controls on the technology; even if there were, the number of nations that manufacture such systems is so large and varied that a potential customer or license producer is almost assured of finding someone willing and able to sell UAV airframes or license production.

Table 2.1

Examples of UAV Airframes That Could Be Converted into Effective and Relatively Inexpensive Cruise Missiles

Specification	Type I Lear R4E Skyeye	Type 2 SDPR VBL-2000
Range (nmi)	650	594
Cruising Speed (knots)	65	74
Payload (kg)	80	30
Gross Weight (kg)	354	150
Wingspan (m)	6	3.3
Length (m)	4	3.25

SOURCE: Kenneth Munson, ed., *Jane's Unmanned Aerial Vehicles and Targets*, Coulsdon, Surrey, England: Jane's Information Group, Issue 0, June 1995; Issue 3, July 1996.

payload, weight, and overall dimensions of each. The Lear is about twice the wingspan and weight of the VBL-2000, and has more than double the payload and slightly longer range. However, both airframes are light enough to be carried on, and launched from, a variety of vehicles—possibly as small as an ordinary pickup truck.

At their typical cruising speed of approximately 70 knots, cruise missiles based on these airframes could reach targets at their maximum range in 8 to 9 hours. Their slow speed and consequent long transit time to their targets may make these proposed weapons seem ridiculous—and very vulnerable to interception by advanced U.S. air defense systems. However, U.S. air defense systems such as the F-15, the Airborne Warning and Control System (AWACS), Patriot, Hawk, and AEGIS were designed during the Cold War to detect and engage high-performance Soviet aircraft and missiles, using the high speed of the Soviet systems to help simplify target processing. That is, the computer-controlled look-down radars ignore, or sort out, slow-moving objects to prevent their data-processing and display capabilities from being overwhelmed by hoards of moving ground vehicles. Most tanks, cars, and trucks move slower than about 80 knots, so these radar systems just ignore potential targets moving slower than that. Some systems do have the capability to detect and track slower targets, but they are able to do so only in very narrow sectors and for short periods before the number of potential targets exceeds their processing capacity.

Surface radars are less affected by ground clutter than are airborne radars, but they suffer from very limited line of sight against low-flying cruise-missile targets. Patriot and AEGIS could track those cruise missiles moving at slow speed, but only if they were close enough to be above the radar's horizon; for cruise missiles flying at 30 to 40 meters, "close" is less than 20 miles. Unless the United States deploys huge numbers of ground-based radars to a future theater, most cruise missiles will go undetected by existing air defense systems.

So, what might at first seem a severe disadvantage—slow speed—is actually one of the strong points of these systems: Existing U.S. air defense systems are *designed specifically* to ignore aircraft traveling at the speed of these proposed cruise missiles. If painted black, provided with effective mufflers, and launched in the late afternoon or early evening, these small, quiet missiles could make their way to targets hundreds of miles away with little chance of visual or audio detection on even the shortest summer night at most latitudes.

To explore how cruise and ballistic missiles similar to those described above could be used by potential U.S. adversaries to degrade, disrupt, or defeat USAF airpower, the following chapter presents an illustrative scenario of a future conflict in the Persian Gulf region.

Chapter Three

ILLUSTRATIVE SCENARIO AND IMPLICATIONS

"To have command of the air means to be in a position to prevent the enemy from flying while retaining the ability to fly oneself."

Giulio Douhet, *Command of the Air,* 1921

The scenario described in this chapter assumes that the USAF reacts to a future crisis in accordance with its current concept for conducting theater air operations or with similar concepts that rely heavily on short-range fighters that require bases within striking range of the missiles we have described. It is not intended to predict when, where, or with whom the USAF is likely to next find itself in combat. Rather, its purpose is to show that even adversaries with relatively moderate means could afford to build and employ a force of effective ballistic and cruise missiles equipped with the technology described in Chapter Two, to delay, disrupt, limit, or defeat USAF combat and airlift operations in a future conflict.

A NEW WAR IN THE GULF

Our scenario is set late in the next decade, circa 2007, when, upon the death of Saddam Hussein, civil war erupts in Iraq. One of Saddam's loyal relatives gains the upper hand and begins to brutally suppress an uprising by the Shi'ite population of southern Iraq. Iran warns that it will not sit idle while its Shi'ite brethren are slaughtered.

Iranian armored units mass on the Iraqi border. The war of words escalates as the United States warns all parties that it will act to restore order in the region if the situation does not improve. Iran warns "non-regional powers" to stay out of the brewing conflict and launches an armored offensive into southern Iraq, rapidly advancing to support the hard-pressed Shi'ites in and around Basra. In response to the crisis, the United States deploys combat aircraft to bases on the Arabian peninsula and begins a massive airlift of troops and supplies into the Persian Gulf region.

The Iranians had a ringside seat for the 1991 Gulf War and have great respect for the capabilities of U.S. airpower. They therefore have anticipated the need to counter USAF power projection. The Iranian military has developed a carefully planned strategy for dealing with U.S. intervention in general and USAF land-based air operations in particular. In broad outlines, the strategy is as follows:

- Allow USAF combat units to deploy into bases within cruise-/ballistic-missile range.
- Allow the USAF to establish strategic airlift operations at a base within cruise-/ballistic-missile range.
- Attack vulnerable bases with a large-scale, surprise missile attack designed to destroy and/or damage as many aircraft as possible and inflict maximum casualties.
- Demonstrate the ability to repeat large-scale attacks on any facility within missile range.
- Avoid the use of chemical and biological weapons, to decrease both negative international reaction and the chances of massive retaliation.

Figure 3.1 shows the locations of postulated USAF deployment bases and the locations and ranges of Iranian missile launches. We chose the following four USAF deployment bases: Dhahran, Riyadh Military, Al Kharj (all in Saudi Arabia), and Doha in Qatar.[1] Whether

---

[1]It is possible that the USAF could deploy into more, or different, bases. However, we chose these four bases for several reasons. Dhahran's enormous parking ramps (see Table 3.1) make it by far the best location for a massive airlift operation into the theater. Al Kharj is where most current USAF activity takes place in Saudi Arabia. Riyadh Military is obviously a Saudi military field and is close to the area of anticipated combat (Basra), but is far enough from Iran to give the impression of greater safety. Doha was used by USAF F-16s during the 1991 Gulf War and has already hosted an Air Expeditionary Force (AEF) deployment.

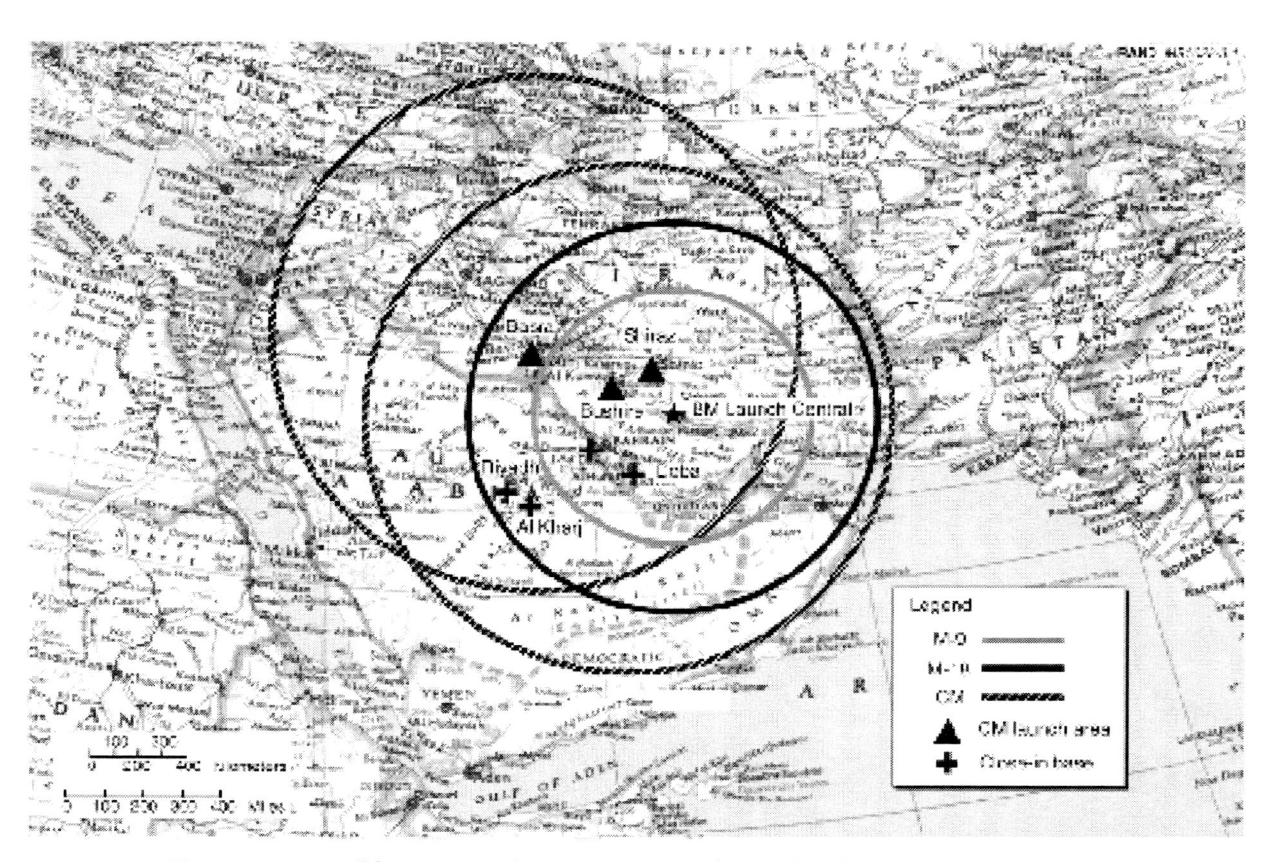

Figure 3.1—Possible USAF Deployment Bases and Postulated Iranian Missile Ranges

USAF units deploy into these bases or others is not important as long as the bases are within about 600 nmi of an Iranian missile-launch point. Although the choice of specific bases is not critical, these bases were used in the past as USAF deployment bases. The critical issue is that short-range aircraft are typically deployed close to the fight in order to reduce their requirement for the limited tanker sorties available. The figure also shows that, with the exception of a few bases in the southwestern corner of the country, most bases in Saudi Arabia, as well as all bases in Bahrain, Qatar, Oman, and the United Arab Emirates (UAE), would be within range of the assumed threat. The Iranians deploy their small, slow cruise missiles on light, truck-mounted launchers, which operate from urban areas, hiding in the clutter. They take advantage of underground parking structures, warehouses, etc., to store and reload their missiles. The cruise-missile units operate from the Shiraz/Bushire area in Iran and from the recently captured Iraqi city of Basra; mobile ballistic-missile operating areas are in the mountains south of Shiraz. All missile-launch areas are protected by mutually supporting SA-10 and SA-12 surface-to-air missile (SAM) launchers.

MISSILE ATTACKS AGAINST USAF ASSETS

With the above scenario providing an operational context, we can now ask, "How effective could missile attacks on USAF airbases be under such circumstances?" We decided to quantify the assessment by determining how many—or how few—missiles it would take to achieve a 90-percent probability of kill (Pk) on aircraft parked in the open at the four postulated USAF operating locations.

Basically, we assumed circular submunition dispersal patterns that would efficiently destroy parked aircraft. We then calculated the total expected lethal area of each submunition warhead by assuming a submunition-dispersal pattern in which half an aircraft dimension lay between the lethal area of each of the submunitions in the pattern. This calculation produces a large lethal area for the warhead while providing a high probability of kill on an aircraft-sized target within the warhead area.[2] The simple approach we used for the calculations is detailed in Appendix A.

The remainder of this section discusses the results of our projected attacks against the four deployment bases.

Attacking Aircraft

Table 3.1 lists the dimensions of the aircraft parking ramps at the four USAF deployment bases. Note the two exceptionally large parking ramps at Dhahran. These two enormous strips of concrete are exactly the sort of facility required for rapid and efficient aerial port of debarkation (APOD) operations, making Dhahran the most likely candidate for the theater strategic airlift hub in this conflict, just as it was during Desert Shield/Storm.

The total area of the parking ramps in the table is over 44 million square feet, the equivalent of almost 1,000 football fields. It can accommodate a huge number of combat aircraft and an intense aerial-port operation. However, the number of GPS-guided, submunition warhead cruise and ballistic missiles required to attack this huge area

Table 3.1

Parking-Ramp Dimensions at Possible USAF Deployment (all dimensions in feet)

Dhahran (likely APOD)	Doha	Riyadh	Al Kharj
9,000 × 900	2,100 × 700	8,400 × 700	6,400 × 1,600
4,200 × 600	1,800 × 900	5,600 × 800	5,000 × 1,000
900 × 900	600 × 600	3,300 × 300	
1,200 × 900	600 × 300		
2,100 × 700			

NOTE: Parking-ramp dimensions are from the National Imagery and Mapping Agency, *DOD Flight Information Publication: High and Low*

[2]Throughout this document, we assume that a hit results in a functional kill, since aircraft are fairly vulnerable to this type of attack and any damage will likely render the aircraft inoperable, at least for the short term. In many cases, the aircraft will be severely damaged from the attack and subsequent secondary explosions.

Altitude Europe, North Africa, and Middle East, St. Louis, Missouri, February 27, 1997.

is surprisingly small. Table 3.2 shows just how few missiles are required for a least-cost attack on the parking ramps described in Table 3.1.

The table lists the number and type of missiles required to attack the parking ramps at the four bases and the tent cities mentioned in the next subsection. We varied the effectiveness of the individual bomblets by repeating the analysis for bomblets with five different assumed lethal radii between 15 and 25 feet. We also assumed that each base is defended by a Patriot or THAAD battery, and that the

Table 3.2

Least-Cost 0.9-Pk Missile Requirements

	Assumed Lethal Radius of 1-Pound Submunition				
Weapons Required to Cover:	15 ft	17.5 ft	20 ft	22.5 ft	25 ft
Parking Ramps					
No. of M-9 Submunitions	46	41	30	29	26
No. of M-18 Submunitions	47	36	30	27	23
No. of CM-1s	36	31	38	23	22
Cost ($M)	$ 151	$ 122	$ 101	$ 90	$ 79
Tent Cities[a]					
No. of M-9 Submunitions	30	24	20	16	14
No. of M-18 Submunitions	30	24	20	16	14
Cost ($M)	$ 90	$ 72	$ 60	$ 48	$ 42
Patriot/THAAD					
No. of CM-1s	12	8	8	8	8
Cost ($M)	$ 3.6	$ 2.4	$ 2.4	$ 2.4	$ 2.4
TOTAL COST ($M)	$ 244.6	$ 196.4	$ 163.4	$ 140.4	$ 123.4

[a]The weapons used to attack tent cities were the same as those used against the aircraft on the ramp. For the tent city, the level of damage would probably not be as high as that for aircraft, since the submunition dispersion was chosen for attack against larger aircraft targets.

fire-control radars for these systems are attacked by cruise missiles.[3] We believe that the dollar values shown are representative of our assumption that the cruise missiles would cost between $200,000 and $300,000 to mass-produce, and that the ballistic missiles would cost approximately $1 to $2 million each.[4,5]

Attacking Tent Cities

Tent cities are another possible target set for ballistic and cruise missiles equipped with submunition warheads. The home-away-from-home for USAF personnel when they deploy to theater operating bases, the tents serve as sleeping and living quarters, dining facilities, showers, and, in many cases, they house important administrative and other functions. As Figure 3.2 shows, these areas are large, typically about 1 square kilometer. The tents, vehicles, streets, etc., are all large enough to be easily visible on commercially available satellite imagery, making detection and targeting fairly simple for countries having the most limited means. Should satellites fail, human-intelligence (HUMINT) sources equipped with GPS could easily provide coordinates close enough to the center of tent cities to ensure a missile hit.

However, to attack the entire area of the typical tent city, so many of the small cruise missiles we have described would be required that the attack would be costly and difficult to coordinate. Cruise missiles could be used in smaller numbers to harass tent-city residents; but

---

[3]This refinement of our operational concept—using the low-flying, slow cruise missiles to attack missile-guidance radars—might be a particularly rewarding tactic, since Patriot and THAAD have only limited capability to engage such targets and eliminating these systems with low, slow cruise missiles could synergistically make ballistic missiles much more effective. Here, *synergistically* refers to the fact that Patriot and THAAD are being designed against the ballistic missiles. However, ballistic missiles carry a much larger warhead than does the cruise missile postulated. Therefore, the most-effective attack may be to use the cruise missiles to damage the anti–ballistic missile (ABM) systems so that the ballistic missiles will be more effective against the aircraft on the ramps.

[4]See Table A.1 in Appendix A for more-detailed data on assumed missile characteristics.

[5]The authors used their best guess for the cost of ballistic missiles, because data were not readily available.

Photo courtesy of U.S. Air Force

Figure 3.2—Tent City at Prince Sultan Air Base (Al Kharj), 1996

any real attempt to cause severe damage to the tents and equipment, personal effects, etc., inside would be far more efficiently undertaken using ballistic missiles.

Approximately 7 to 15 M-9/M-18 missiles with submunition warheads could cover the typical 1-square-kilometer tent city. Obviously easier to plan and execute than an attack with scores of small cruise missiles, such an attack has a potential disadvantage when compared with a cruise-missile attack. Base personnel will almost certainly receive warning of an incoming ballistic-missile attack and take shelter in trenches, bunkers, or other hardened facilities, so that personnel casualties will likely be fairly light. However, shredding or setting fire to most of the tents, vehicles, and the equipment and personal effects inside would have severe consequences for the base's ability to generate sorties. Even if most of the personnel at a base were unharmed in the attacks, their immediate efforts would almost certainly be devoted to extinguishing fires and later salvaging whatever aircraft and equipment they could from the wreckage, rather than generating effective combat sorties.

With most shelter, clothing, and many work centers shredded or destroyed, cleanup, clearly, would be the next order of business. Salvage and recovery operations would probably be directed toward evacuating the base and moving the unit's surviving equipment to a location beyond the reach of enemy missiles. There, a more complete assessment and recovery could begin, free from the threat of further attack.[6]

However, this process would probably be long, frustrating, and interrupted often by further missile attacks. Surviving combat aircraft could probably leave the base once they had been checked for damage and a foreign-object-free corridor was cleared across the parking ramp so that they could taxi to the runway. However, ground transportation would have to be enlisted to move surviving ground equipment to the new operating location or to an airport outside of the missile range where airlift operations could proceed without putting valuable cargo aircraft at risk.

IN SUM

This chapter has outlined how a hostile power could use faily simple (relative to the Tomahawk), readily available missile and UAV technology to launch attacks against USAF operations at theater airbases. We have shown that, for about $1 billion, an adversary could attack four missile-defense radars once, four tent cities once, and all parking ramps between 6 and 12 times each. These attacks have the potential to be so destructive to equipment and disruptive to sortie-generation operations that, unless steps are taken to diminish the effectiveness of these systems, they could force the USAF to abandon bases within reach of enemy missiles. Chapter Four outlines several alternatives available to the USAF to improve the survivability of theater airbase operations in the face of a missile threat like the one described above. It also examines some of the potential costs and impacts on operations of these options. Chapter

[6]For the adversary to replace some of the bomblets with mines could significantly enhance the effectiveness of an airbase attack. Returning the base to operational status would be delayed, since the mines would need to be cleared before the cleanup effort could get under way.

Five examines several operational concepts that would enable the USAF to operate effectively from beyond the effective range of the threat.

Chapter Four

DEFENSIVE RESPONSES TO AN ENEMY-MISSILE THREAT

In this chapter, we discuss two different sets of defensive options available to the USAF to counter submunition-carrying missiles. The first set consists of several passive defenses the USAF could adopt to help protect forward-deployed assets; the second set consists of active-defense measures.

Passive defenses are defenses employed by the defender to reduce the impact of attacks by enhancing its own ability to absorb and recover from attacks. As the name implies, once constructed, passive defenses just sit there. They have taken many forms throughout human history, from wood-and-mud stockades around a group of huts, to the walls of medieval cities, to the barbed wire and trenches of World War I. They are remarkably effective because they are usually simple, reliable, require little or no operator effort to use, and the defender can construct obstacles and protective shelters at leisure before a conflict starts. In contrast, the adversary must wait until the war has started to breach or dismantle these defenses and must do so under fire.

Active defenses have traditionally required an operator—whether an archer on a castle wall, a machine gunner in a foxhole, or a surface-to-air missile battery operator—to detect, track, identify, engage, and hit an attacker. They tend to be more complex than passive defenses and require considerable operator skill and training. However, they have an advantage: Instead of delaying or deflecting attacks as passive defenses do, they have the potential to cause substantial damage to enemy equipment and inflict serious casualties.

As is often the case in warfare, active and passive defenses, when combined, tend to produce far better results than either could alone. The archer is much more effective from the top of the castle wall, where he has a commanding field of fire; the machine gunner is less vulnerable and more effective from the protection of a trench behind a barbed wire entanglement than on flat terrain. Similarly, the passive and active defense measures described in the following sections should be thought of as mutually reinforcing measures to reduce the effectiveness of an enemy missile attack rather than as competing, stand-alone solutions to the base-vulnerability problem.

PASSIVE DEFENSES

In the following subsections, we present four possible passive defenses for the postulated scenario: fixed hardened shelters, deployable shelters, dispersed aircraft parking, and dispersed operations.

Fixed Hardened Shelters

During the later years of the Cold War, NATO built hundreds of hardened concrete shelters to counter possible Soviet aircraft and missile attack, and the Soviets built similar shelters at their opposing bases in Eastern Europe. The United States and its allies have constructed some shelters in South Korea and throughout the Middle East. However, existing facilities in many areas are inadequate to accommodate the number of combat aircraft the USAF would likely deploy to successfully execute its planned air operations. So, without additional shelters, USAF combat aircraft would have to be parked in the open as they were at many bases during the 1991 Gulf War.

Hardened concrete shelters would provide excellent protection against the small submunitions that we postulate future adversaries attempting to shower on USAF theater airbases. In addition, they provide a measure of protection against terrorist and special operations force (SOF) attacks, and help protect aircraft from contamination by gross chemical and biological agents.

However, they do have at least four serious disadvantages:

- They are expensive. A shelter large enough for a single fighter-size aircraft like the one shown in Figure 4.1 costs approximately $4 million.[1] Enough hardened shelters to protect the aircraft of five 72-aircraft fighter wings would cost over $1.4 billion. In addition to protecting its combat aircraft, the USAF would also need to build sufficient hardened facilities to protect vulnerable aircraft-generation equipment, spare parts, avionics-repair equipment, command and control facilities, operations, and so on, greatly adding to the price tag.
- Sheltering aircraft, maintenance activities, and other important functions does nothing to reduce the vulnerability of tent cities to missile attacks. Providing hardened living and work facilities for the thousands of people required to keep an Air Force fighter wing operational would be extraordinarily expensive. NATO air-bases in Europe were moving toward a protective posture for all personnel. However, the expense was so large that, even during the peak defense-spending years of the mid-1980s, NATO air forces could not afford to build enough hardened shelters to protect all of their personnel.
- Because of the size and shape of many of the USAF's large, high-value aircraft (Joint Surveillance, Targeting, and Reconnaissance System [JSTARS], Rivet Joint, AWACS, strategic airlifters, etc.), no attempt has been made to design or build hardened shelters large enough to accommodate them. Unless the design challenges of building hardened structures large enough to accommodate large aircraft can be overcome, adopting a hardening strategy that protects only fighter-size aircraft while leaving high-value assets at risk is a partial solution at best.
- This is possibly the most telling disadvantage of fixed shelters: If they are to be of any use at all, the hardened shelters must be in the right place. The expense and time required to construct adequate hardened shelters requires that Air Force planners anticipate correctly, years in advance, which bases will be used in a fu-

---

[1] Personal conversation, May 28, 1997, with LtCol Rick Parker, AF/CE.

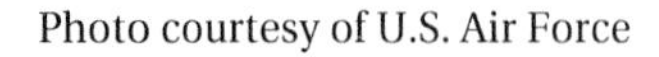

Figure 4.1—F-117 Fighter Being Towed into a Concrete Hardened Aircraft Shelter

ture war. If troops are forced to fight in an unanticipated region, or from bases different from those the Air Force had planned on using in a region in which it had planned to fight, the shelter investment will have been wasted, and U.S. aircraft will still be vulnerable to enemy missile attack.

Deployable Shelters

The level of protection that fixed concrete shelters provide is far in excess of that required to defeat submunitions. It may be possible to design and build transportable shelters that can adequately protect against submunitions.

Deployable shelters, made of Kevlar over a lightweight aluminum frame, could be prepositioned at central locations in anticipated theaters to minimize the total inventory required, then be moved rapidly to any actual USAF deployment locations. It might also be possible to design shelters large enough to provide some protection to airlifters and Civilian Reserve Airlift Fleet (CRAF) aircraft conducting strategic airlift operations.

However, providing the necessary level of ballistic protection in a lightweight package is not easy. We calculated that an F-15–size fighter would require a shelter with minimum dimensions of 15 × 22 × 7 meters, which would allow just enough space for the aircraft. A larger shelter would probably be required to allow for normal maintenance activities. Assuming bomblet fragments hitting the shelter walls have no more energy than a .44 Magnum pistol round, 9-millimeter (mm)-thick LASTEC 30 (i.e., bullet-resistant-material) walls could provide adequate protection with a weight of 10 kilograms (kg) per square meter. To protect against any bomblets that landed directly on top of the shelter, the ceiling would have to be reinforced with additional armor. Assuming the "high-energy" fragments from these direct hits possess less energy than a 7.62 × 51 mm NATO round, the additional protection would increase the ceiling weight to 12 kg per square meter.[2] Allowing for additional weight for a supporting frame, we calculated that, using a typical bullet-resistant material, deployable shelters for 72 F-15–size aircraft alone would require 22 C-17 loads.[3]

As with fixed shelters, protecting the aircraft is only the first step. Whereas some equipment and stores may be quite resistant to submunitions (heavy machine tools in German factories proved remarkably resistant to bombing during World War II), many of the items needed to keep a modern fighter wing producing sorties at a high rate are not as robust. Spare engines, electronic test equipment, spare avionics, and the trained personnel who use and service this equipment and the aircraft themselves must also be protected if a

---

[2]Ian V. Hogg, ed., *Jane's Security and CO-IN Equipment*, Coulsdon, Surrey, England: Jane's Information Group, 1991–1992.

[3]Since this material is dense, the C-17 will reach maximum load before reaching maximum volume. For example, a 9-mm-thick sheet of LASTEC 30 weighs a little over 10 kg per square meter, which works out to over 1,100 kg per cubic meter. According to *Jane's All the World's Aircraft* (Jackson, 1997–1998), the C-17 has an internal volume of nearly 592 cubic meters. Loading the aircraft to its full volume, a C-17 with LASTEC 30 would weigh over 650,000 kg—an order of magnitude greater than the typical payload. Therefore, the C-17 will reach maximum weight long before reaching maximum volume, even considering packing inefficiencies and the volume of the aluminum frame.

base is to continue to generate combat sorties following a missile attack.

The benefits of deployable shelters for increasing the number of missiles required to destroy parked aircraft are shown in Figure 4.2.[4] We used the same procedure presented in Chapter Three to compute the number of missiles required to cover the parking areas at Dhahran for the "no-shelter" cases. In doing the M-9/M-18 calculations (for the no-shelter case), we assumed a submunition warhead (20-foot lethal radius for each submunition). The other cases assume unitary warheads.

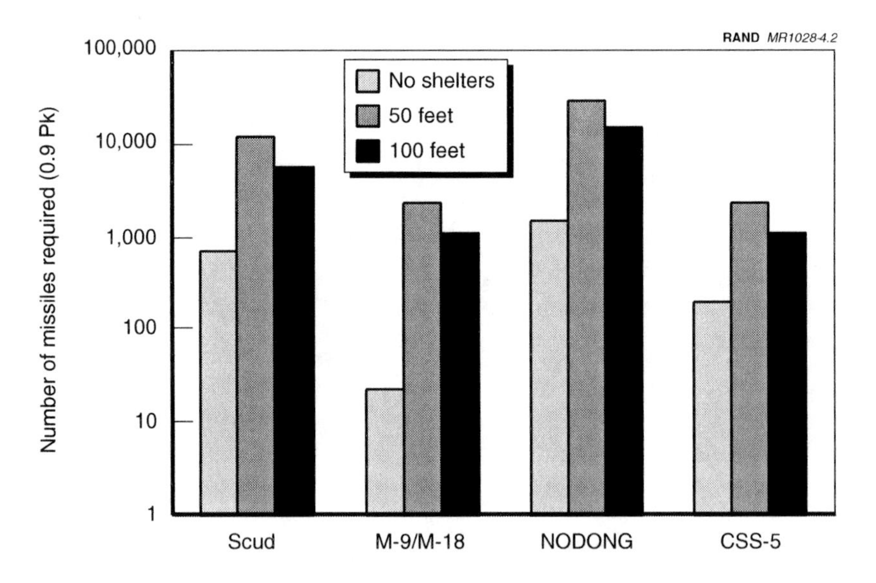

Figure 4.2—Missiles Required to Attack Dhahran Parking Ramps With and Without Shelters

[4]Note that the USAF is fully committed to becoming an expeditionary force that would emphasize its ability to rapidly deploy anywhere in the world. Therefore, the USAF is unlikely to limit its deployment options by building infrastructure such as shelters or increased parking areas (described in the following subsection). Moreover, the expeditionary concept will not easily accommodate the deployable shelters we have described, because they dramatically increase the airlift assets required to deploy a given force.

To calculate the number of missiles required to destroy the deployable shelters, we assumed a 72-aircraft wing deployed to Dhahran with an equal number of shelters. The sheltered aircraft are attacked using missiles with unitary warheads. In the two cases presented, we varied the distance within which the unitary warhead must strike the center of the deployable shelter to register a kill. In the first case, the weapon must land within 50 feet of the center of the shelter—essentially landing within about 20 feet of a shelter wall—to register a kill. In the second case, the weapon must land within 100 feet of the center of the shelter—about 70 feet from a shelter wall—to kill the aircraft inside. The numbers presented here assume that the adversary has perfect targeting information for each shelter on the ramp (TLE = 0). Note that the scale is logarithmic to facilitate reading the chart. In all calculations throughout this document, we assume a 90-percent probability of kill and the same reliability factor for each weapon.

Dispersed Aircraft Parking

Another possible way to complicate enemy attempts to attack USAF aircraft parked in the open is to disperse those aircraft over a much larger parking area. Figure 4.3 illustrates one way to achieve that dispersion: taxiways extending outward radially from a base the USAF anticipates using in a theater conflict. These long taxiways would be constructed with many hardstands—far more than the number of aircraft operating from the base. The length of the taxiways and their radial geometry would maximize the number of missiles required to attack the base.

The number of missiles required to attack a given base is a function of the distance between contiguous hardstands. Figure 4.4 shows how the number of missiles required to attack 54 hardstands increases with increased hardstand spacing. If we assume that an adversary has accurate data on where aircraft are parked on the airbase, the number of weapons required to attack those aircraft successfully increases with the distance between hardstands, until that distance is so great that each aircraft must be targeted individually. For missiles with the characteristics we have described, increasing aircraft spacing beyond 750 feet does not increase the number of missiles required to attack them: When using accurate missiles, an

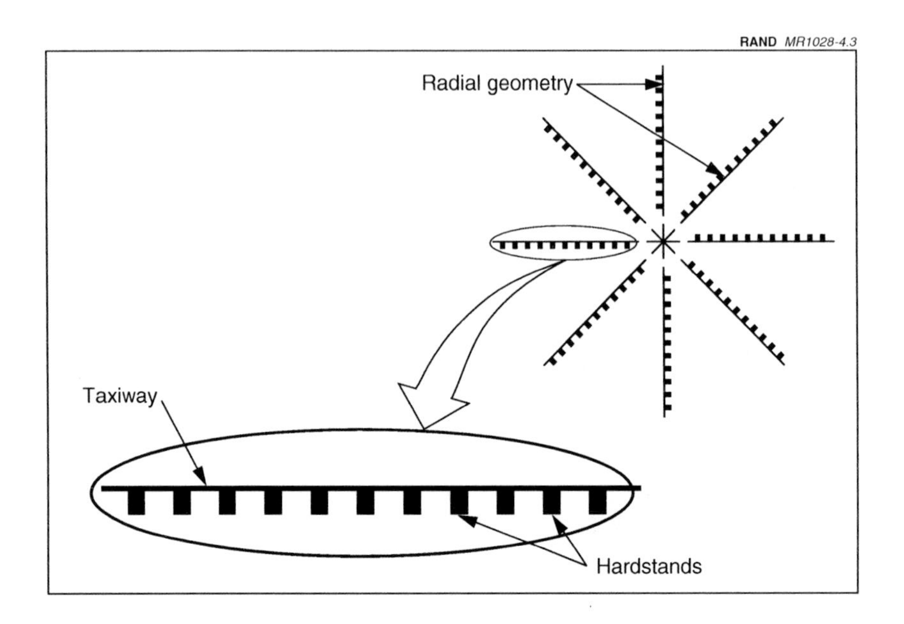

Figure 4.3—Additional Aircraft Parking-Ramp Geometry of Hardstands Built Outward Radially from the Base to Maximize Weapon Requirements

adversary could simply target each hardstand, which would be the maximum number required.

One possible alternative to building additional parking ramps like those in Figure 4.3 is to simply space aircraft 750 feet apart on existing parking ramps. However, giving each aircraft its own 750 × 750 foot area of existing ramp space places extreme limits on the number of aircraft the USAF could operate from existing facilities. For example, even the huge—7.2-million-square-foot—parking ramp at Dhahran could accommodate only 12 to 15 aircraft parked in this fashion. Therefore, it may be impractical to achieve the maximum benefit of dispersion using existing facilities. Radial taxiways like those in Figure 4.3 minimize the amount of new concrete required.

This approach has several limitations:

- The geography and land-use patterns around potential operating bases may make it difficult or impossible to adopt this approach.

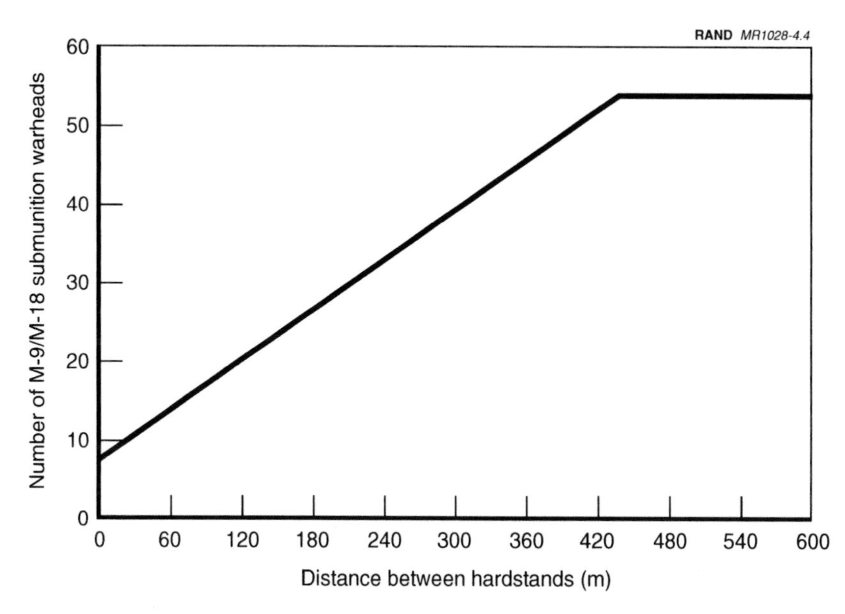

Figure 4.4—Weapons Needed to Attack a Fighter Wing As a Function of Distance Between Parking Spots

- It is not clear that it would be cheaper to build long taxiways to unprotected hardstands than to build shelters to provide protection against submunition warheads.
- As with hardened shelters, correctly anticipating from which bases the USAF will need to operate in a future conflict must be done far enough in advance to complete the dispersal taxiways.
- Even an adversary that has access to overhead targeting data or HUMINT sources might still believe that targeting each $50-million U.S. tactical aircraft with individual $2-million ballistic missiles is a good exchange.
- This parking scheme would vastly increase the area USAF security personnel must patrol and monitor, and it does not address the vulnerability of tent cities. With this increase in base area comes a likely increase in the probability of a successful terrorist or SOF attack.

A potentially more-flexible and inexpensive variation on the dispersal option would be to lay metal-mat parking areas around USAF theater operating bases after U.S. forces deploy. These temporary ramps could be moved periodically to complicate an adversary's targeting problem, and materials could be stockpiled in central locations in likely theaters to minimize the logistics burden and time for construction.

However, these temporary ramps present at least three potential disadvantages:

- If an adversary knows where the additional temporary ramps are, or can get this information faster than U.S. forces can move the ramps, only a few additional missiles would be required to attack the temporary ramps, unless they are arranged in the radial geometry described above. For example, attacking two 1.5-million-square-foot parking ramps would require only two additional M-9 or M-18 missiles.
- The temporary parking ramps and taxiways connecting them to existing base infrastructure would pose a serious foreign-object-damage (FOD) threat to most existing USAF combat aircraft. Unlike Russian combat aircraft, such as the MiG-29, USAF fighters were designed to operate from well-swept concrete and asphalt surfaces and have no built-in protective intake covers to prevent their engines from ingesting dirt, sand, rocks, and other debris likely to be kicked up as aircraft taxi over the flexible metal mats of the temporary parking areas. F-16s, with their low-mounted jet intakes, would be especially vulnerable to FOD under these conditions. This particular disadvantage could be addressed by providing removable intake screens for existing USAF combat aircraft. They could be removed or installed in the arming areas before and after each flight. Future combat aircraft could be designed with anti–FOD-intake protection systems that would allow them to operate under more-austere conditions than current USAF fighters do. Alternatively, aircraft could be towed to and from the metal-mat parking areas.
- The metal-mat-parking-area solution does nothing to decrease the vulnerability of tent cities to missile attack.

Dispersed Operations

Another alternative for dealing with the missile threat we have posed is to abandon operations from fixed theater operating bases and disperse operations to a large number of small highway strips capable of handling 5 to 10 aircraft. An advantage of this option is that it dramatically complicates the attacker's targeting problem by dispersing what was a single large, lucrative target into many much smaller pieces that, individually, are much less vulnerable.

However, dispersed operations would almost certainly complicate and reduce the efficiency of USAF combat operations. Aircraft operations achieve significant efficiency gains from large economies of scale. For example, aircraft egress systems are so reliable that a wing needs only 7 to 10 ejection-seat specialists to keep all of its aircraft-egress systems working. However, if the wing is dispersed to 10 or 15 separate operating locations, not enough specialists will be available for each location to have its own. In addition, many jobs have a complexity that places special demands on a team of specialists. They might take more than one specialist to complete, one individual's work needs to be inspected and signed off by a second to ensure that critical systems are properly repaired, and specialists need to rest. Consequently, the wing probably has only two or three egress teams available to cover the work at 10 to 15 dispersed locations. Similar arguments can be made for fuel, hydraulic, propulsion, avionics, and other specialists.

To support dispersed operations, a wing will either need to increase its maintenance personnel in many specialties or be able to rapidly transport specialists from one dispersed operating location to another. Delays associated with the need to transport spare parts, especially costly or large items such as spare engines, to the dispersed locations are other possible sources of inefficiency that would decrease sortie generation. Requiring aircraft to land with sufficient fuel reserves to take off and immediately refuel from a tanker operating from a base beyond missile range might solve the fuel-delivery problem at the expense of some combat radius,[5] but delivering the

[5]If the aircraft needs to land (after a mission) with enough fuel to take off (no ground refueling) for the next mission, it cannot go as far on the first mission; therefore, in the aggregate, combat radius is reduced.

right quantities and types of munitions to the dispersed locations would probably be as difficult as delivering the right mix of spare parts and maintenance specialists.

Another problem area would be command-and-control and mission planning. It might be possible to create an electronic command and control system that would allow dispersed aircrews access to the target, intelligence, and other information they need for planning their own missions. However, no such system currently exists, and it would probably be expensive to develop and deploy. Such a system would probably also need to have some sort of remote mission-briefing capability, so that crews operating from separate locations could plan and fly missions together if the USAF wants to retain scheduling flexibility and the capability to put together strike packages exceeding the number of aircraft at a typical dispersed location.

Force protection under this type of operational concept would be an extreme challenge. Relatively small, widely scattered groups of USAF maintenance personnel and pilots would be exceedingly inviting targets for terrorist or SOF attacks. To protect each operating location from attack by short-range air defense (SHORAD), mortar, or anti-tank guided-missile systems, an area extending out at least 5 km from the edges of the dispersed location must be monitored. This area would be almost as large as the security zone surrounding a single large main operating base, since the dispersed-location runways would need to be roughly the same length as the runway at a single concentrated base. Dispersing operations to 10 or 15 locations would thus increase the area that wing security personnel must monitor and control, by something like 10 to 15 times. Protecting such a large area would require far more security personnel and resources than USAF wings currently possess.

Finally, an adversary with overhead imagery or HUMINT sources that could identify the dispersed operating locations could still target those locations with the types of missiles we have described. Effectiveness against the missile threat would require that dispersed operating locations be capable of rapidly moving between several alternative operating sites, which would further reduce sortie-generation capability.

Dispersed operations might prove to be an effective passive defense against enemy missile attacks. And, unlike the other passive-defense options discussed above, dispersed operations decrease the vulnerability of both aircraft parking areas and tent cities to missile attack. However, to retain current USAF sortie-generation capabilities, this operational concept would probably require a huge increase in the number of maintenance, security, and support personnel. In addition, it would require the USAF to devise completely new concepts of command and control and logistics, which would probably require expensive new computer and communications systems.

Alternative Dispersed-Basing Concept

One possible modification of the dispersed-basing concept that could address some of the concerns raised here is for each wing to maintain a "full-service" base, outside of enemy missile range, with complete maintenance capability; and a forward refueling/re-arming base closer to the target area and within missile range. Aircraft could begin each day's activities at the full-service main operating base (MOB), fly a mission, and recover at the more-austere and less-capable forward operating base (FOB). To minimize aircraft exposure time at the FOB, inspections and some servicing might be abbreviated or deferred. The aircraft and crew could fly one or more sorties from the FOB before ending their day by flying back to the MOB, where any deferred inspections, servicing, and minor maintenance items could be attended to, free of the threat of missile attack. A wing using this basing system would have higher sortie rates than would a wing operating solely from a MOB outside missile range, as a result of reduced transit times to and from the target for any sorties flown from the FOB. This system would also expose fewer ground crew to missile attack than would the system for a wing attempting to operate from a MOB inside enemy missile range. However, these advantages come at the cost of many of the above-described logistics, command-and-control, force-protection, and personnel problems associated with dispersed operations.

Concluding Remarks

This discussion of passive defenses may leave the reader feeling that there are no practicable passive-defense options against the sort of missile attacks we have postulated. Not so. Aircraft shelters, either fixed or deployable, with sufficient ballistic protection against submunitions, appear to hold considerable promise. Dispersing aircraft at existing bases or to alternative locations could also work.

However, all of these options come at a substantial cost. Fixed shelters are expensive and may not be in the right place. Deployable shelters are heavy, imposing a large, new logistics burden for deploying air wings. Dispersing aircraft at existing bases with permanent new parking facilities could also work, but may be as costly as building shelters; it shares the disadvantage of being nondeployable. Building temporary parking ramps may be feasible, but would require the USAF to develop equipment and procedures to operate its existing and future combat aircraft from much more austere ground facilities than they are currently capable of using. Finally, dispersed operating locations could help counter the missile threat, but may create new security problems and impose huge new personnel costs.

The USAF may need to adopt a mix of passive-defense options. For example, it could work with allied governments to build additional fixed hardened shelters in theaters where the United States has vital interests and anticipates conflict while acquiring enough deployable shelters for two or three wings, to hedge against a conflict in an unanticipated area. The important point is that the USAF should begin seriously examining which passive-defense options it might adopt to complement the active-defense options discussed in the next section.

ACTIVE DEFENSES

The active defenses discussed in this section are to detect, identify, track, engage, and damage or destroy the two types of offensive systems an adversary could use to attack USAF bases: ballistic missiles and cruise missiles. The ballistic-missile defense problem has been studied in great detail by others, and proposals for dealing with it range from counterforce strikes and patrols to boost-phase intercept using missiles or directed-energy weapons to anti-missile systems such as THAAD and the Navy's Upper Tier. We do not discuss the merits of these various proposals or their feasibility here. However, we note that it appears that these defenses will cost billions of dollars to deploy. In this section, we impart some initial thoughts on the types of measures the U.S. military could take to counter the low, slow, small cruise-missile threat we have described.

As discussed in Chapter Two, most existing U.S. air defense systems were not designed to detect, track, or engage threat systems with the speed and size of our postulated cruise missiles. Whereas the ballistic-missile threat has multibillion-dollar research programs to develop active defenses to counter it, there are no such programs for cruise missiles. The lack of such programs is not surprising, for two reasons. First, it is not clear that the threat exists, since no one has yet used such a system in combat.[6] Second, the nature of the threat probably does not require such a large and expensive research program, because the problems involved in detecting, tracking, and engaging small, slow cruise missiles are much more similar to those associated with intercepting high-performance aircraft than are the problems associated with ballistic-missile defense.

Below, we examine some relatively inexpensive active-defense options available to the USAF for countering low, slow cruise missiles in the short, medium, and long term.

Short-Term Active Defenses

GPS Jamming. Jamming GPS-guided ballistic and cruise missiles could be an effective counter. At Earth's surface, the strength of the GPS signal is weak (5×10^{-17} watts).[7] A low-power transmitter could potentially jam GPS signals over a large area. For example, a 1-watt noise jammer could jam the signal for a range of 4.5 km against a receiver with no anti-jam countermeasures.[8] The overall effectiveness of GPS jamming depends on a variety of factors and differs markedly when such jamming is used against cruise missiles than when used

---

[6]The same could have been said about aerial torpedoes and 16-inch battleship gun projectiles modified for use by Japanese naval aircraft against U.S. battleships at Pearl Harbor prior to December 7, 1941. For a description of the extensive and unique weapon modifications and training program adopted by the Japanese Navy to ensure the success of the Pearl Harbor attack, see Gordon W. Prangue, *At Dawn We Slept*, Norwalk, Conn.: The Easton Press, 1988, Chapter 19.

[7]Gerald Frost, *Operational Issues for GPS-Aided Precision Guided Weapons*, Santa Monica, Calif.: RAND, MR-242-AF, 1994, p. 36.

[8]Frost, 1994, p. 29.

against ballistic missiles.[9] These factors include terrain and system specifics, but a few generalizations can be made with regard to jamming effectiveness.[10]

Two methods are conceivable to jam the GPS guidance system of a missile: noise jamming and "smart" jamming. Given that it could be difficult to jam the GPS over the launching country, we focus on target-area jamming in the following discussion.

Noise jamming is intended to drown the GPS signal reaching the missile in a sea of static-like noise. A missile entering an area in which the noise is stronger than the GPS signal would be forced to rely on its inertial navigation system for guidance. As noted in previous RAND research, the accuracy of a gyro can be represented by a drift rate that is a function of time.[11] Since the flight times of cruise missiles are significantly greater than those of ballistic missiles, noise jamming will have a much greater effect on cruise missiles. The effectiveness of this jamming technique will depend on the countermeasures incorporated into the missile.

Although noise jamming offers some hope of decreasing the accuracy and lethality of cruise missiles, this benefit comes at a cost. The possible deleterious effects of noise jamming on friendly forces must also be considered.

"Smart" jamming seeks to drag the weapon off target by providing bogus GPS input signals. However, this approach is technically challenging, since the jammer must know both what satellites are being observed by the missile and the intended aimpoint to accu-

---

[9]This discussion assumes that the missile systems have a GPS–INS navigation system. Such systems must remain within a jammer's footprint long enough for the INS to develop significant drift errors. An INS with a 0.1-degree-per-hour drift rate deviates from course at about 10 feet per minute. To generate an average miss distance of 1,000 feet (enough to miss most, but not all, ramps), the INS must go without updates for 100 minutes. If a cruise missile is traveling at 70 knots, it would have to be in the effective footprint of a jamming system for at least the final 117 nmi of its flight to miss by 1,000 feet on average. For systems equipped with only GPS guidance, limited jamming around potential targets would be an effective countermeasure, since the missiles would be completely dependent on GPS information.

[10]For an in-depth discussion of the issues associated with GPS jamming, see Frost, 1994.

[11]See Frost and Lachow, 1996a, p. 11.

rately provide the fraudulent signal. In some cases, knowing the target will be unnecessary; in others, the fraudulent signal could cause the missile to impact in an undesired location (e.g., in a city).

Lethal Defenses. The slow speed, approximately 70 knots, of the cruise missiles we have described is potentially both their biggest advantage and their biggest disadvantage. It allows them to evade existing U.S. airborne air defense systems by being sorted out as clutter or ground traffic. However, when combined with predictable, computer-controlled flight paths, their slow speed and small size make them near-ideal targets for optically aimed medium and heavy machine guns. This weakness could allow the USAF to quickly employ simple, existing equipment to good effect against these systems. Radar-guided guns could also be highly effective.

One approach would involve deploying small machine-gun teams consisting of a gunner and a loader/lookout around bases. These teams could wear night-vision goggles to detect, track, and engage small, slow cruise missiles. The team's fields of view and fire could be increased if they were placed in 20- to 30-meter-tall towers surrounding the base, from which well-trained gunners should be able to effectively engage targets out to at least 300 meters.[12] Figure 4.5 shows how less than a dozen tower-top teams could provide an interlocking defense for a 1-square-km tent city and two parking ramps. The gunners would probably not be able to destroy all incoming missiles; therefore, some form of shelter for aircraft and personnel would still be necessary.

A gun team detecting a 70-knot cruise missile 500 meters from the edge of the tent city could provide base personnel with only about 15 seconds' warning. Another ring of approximately 30 towers 1 km beyond the first set of towers could provide an additional layer of defense to cause further attrition of attacking cruise missiles, and could increase warning time to between 45 seconds and 1 minute.

---

[12]The cruise missiles we are discussing have lengths and wingspans in the 3–4-meter range. Although the machine guns mentioned, especially 0.50-caliber weapons, can be effective at much more than 300 meters, it would probably be difficult for teams to acquire such small targets beyond about 500 meters. Allowing 5 to 7 seconds for the team to communicate and bring its weapon to bear, we calculate that a target moving at 70 knots (36 meters per second) would come within about 275 to 325 meters of the team.

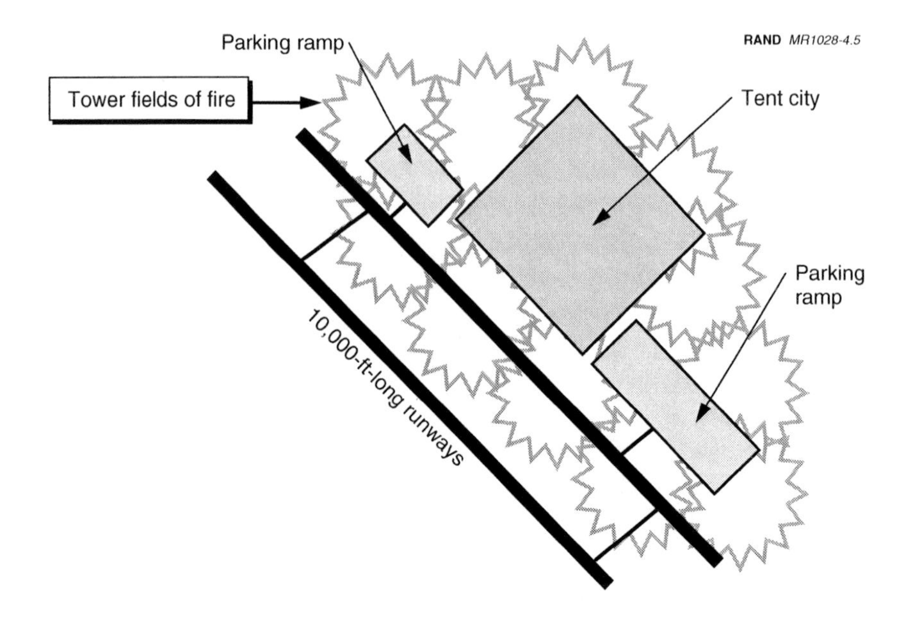

Figure 4.5—Fields of Fire Around a Theater Operating Base, from an Anti–Cruise Missile Gun Tower

Manning the dozen close-in towers with three 8-hour, 2-person shifts would require 72 personnel.

Manning the full 42-tower 2-layer defense in the same way would require about 250 personnel. Machine-gun teams might be supplemented by Stinger missile teams, but the small size of the cruise missiles might make it difficult to detect them at anything like the 3-nmi range of the Stinger. In addition, the expense and slow rate of fire of Stinger teams would probably make them a much less attractive active-defense option than machine gunners.

Medium- and Long-Term Active Defenses

Over the medium and long term, more-sophisticated defenses could be developed that engage small, slow cruise missiles farther away from their targets, provide more warning time, and are more efficient and more effective than the simple gun teams described above. In the medium term, existing systems such as AWACS, JSTARS, Patriot, or the Navy's Phalanx anti-missile gun system might be modified to detect, track, or engage low, slow cruise missiles before they reach their intended targets. These modifications might take the form of new computer hardware or software to allow systems to track targets moving in the 60–90-knot range. Ground-based systems such as Patriot would suffer from line-of-sight limitations, even if modified to track small, slow cruise missiles.

Modifying JSTARS to track such targets seems particularly attractive. From the start, this system was designed to detect and track thousands of slow-moving ground targets and fuse these data with known roads. If JSTARS is sensitive enough to detect small airborne targets like the cruise missiles we have described, then any targets moving toward U.S. or allied forces at 60 to 90 knots *not* on known hard-surface roads would be prime candidates for identification as low, slow cruise missiles. The target information could then be passed on to new gun or missile systems designed to engage cruise missiles.

Over time it might also prove possible to field modified versions of existing systems, such as a land-based Phalanx derivative or the Avenger missile/gun system, that would enable low, slow cruise missiles to be acquired, tracked, and engaged effectively.

All of these proposed long-term adaptations to this postulated threat might require extensive modification to existing hardware and software, and would necessitate extensive testing and development efforts.

In the long-run, more-effective countermeasures can be developed. These might include air defense systems designed specifically to counter the slow-cruise-missile threat. These systems might employ alternative methods of detecting and tracking their slow-moving targets, such as networks of acoustic sensors, passive millimeter-wave radar, or thermal imaging systems.

This chapter has discussed a number of passive- and active-defense options for countering the threat to USAF theater operating bases posed by cruise and ballistic missiles with submunition warheads. The next chapter explores another option for dealing with the missile threat: operating from stand-off bases, bases beyond the effective range of threat systems.

Chapter Five

STAND-OFF OPTIONS

Adopting a stand-off operational concept has benefits for countering the emergence of a cruise and/or ballistic missile threat to USAF operations from close-in airbases. This chapter explores the benefits and costs of the stand-off concept. It has two main sections. The first section deals with a fairly short-term stand-off operational concept the USAF could adopt with its current and projected combat-aircraft inventory. The second section examines the potential for a long-term stand-off strategy for operating and employing U.S. land-based airpower; this strategy would require aircraft with very different characteristics from anything the USAF currently owns or plans to acquire.

SHORT-TERM STAND-OFF OPTIONS

One of the simplest options for dealing with the cruise-missile and ballistic-missile threat is to simply avoid it by operating from bases outside the range of enemy missiles. Figure 5.1 shows the four alternative deployment bases the USAF could use in our hypothetical conflict with Iran: King Abdul Aziz, Taif, and Khamis Mushait in Saudi Arabia, and Thumrait in Oman. All of these bases are outside our proposed cruise-missile range and beyond the reach of ballistic missiles in the Iranian inventory in this scenario, with the possible exception of a few dozen CSS-5s, which cannot conduct militarily significant conventional warhead attacks against the stand-off bases because of their small numbers.

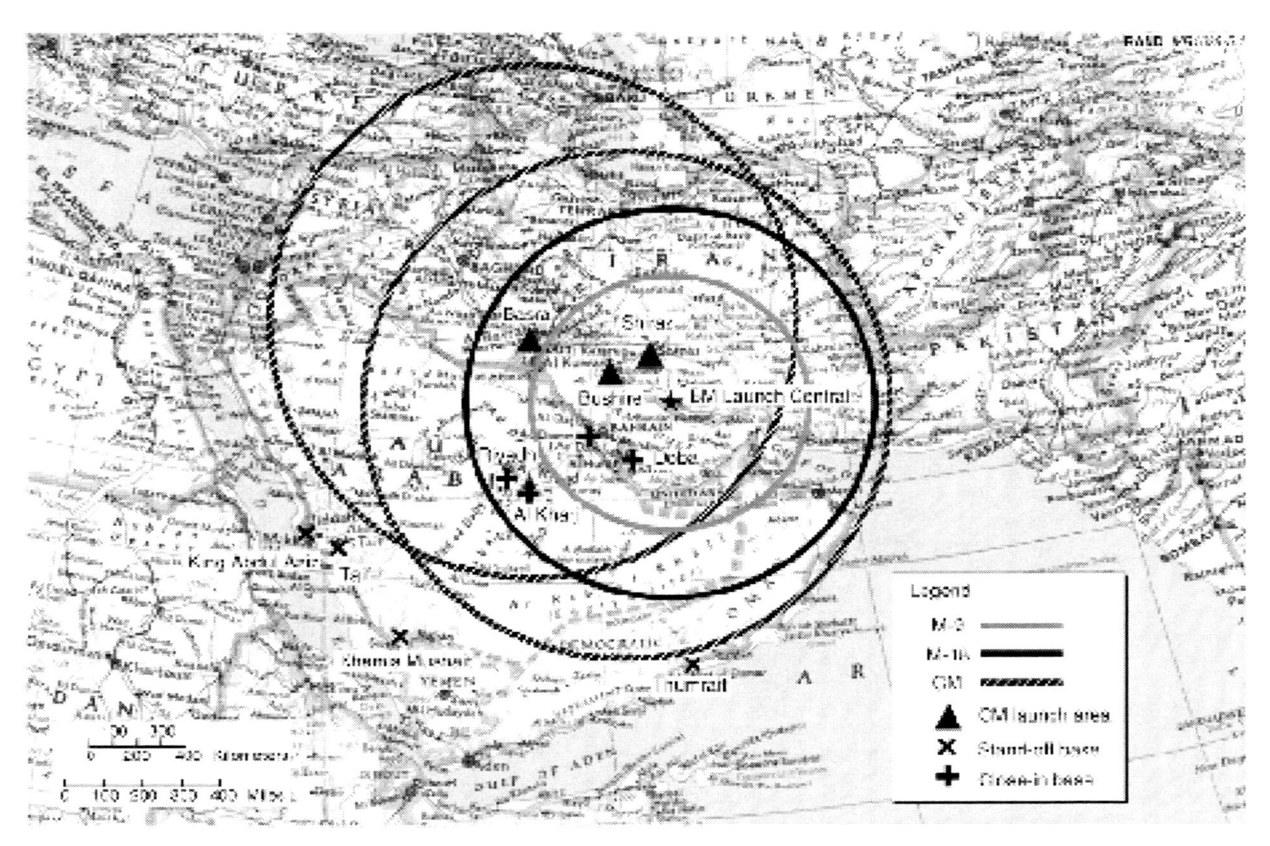

Figure 5.1—Stand-Off and Close-In Deployment Bases with Possible Iranian Missile Ranges

Compared with the four close-in bases described in Chapter Three, the four stand-off bases are much farther from the Basra area—an average of 749 versus 362 nmi. At an average cruise speed of 500 knots, aircraft operating from the stand-off bases will have to fly an average of 3 hours round-trip to targets in the Basra area, whereas their average round-trip time from the closer bases would have been less than 1.5 hours. This increased transit time decreases the number of sorties an aircraft can fly in a given day, for several reasons:

- The aircraft spend more time in transit to and from the target, reducing the maximum number of trips possible in a given amount of time.
- The increased flight time increases wear and tear on some aircraft systems, which, in turn, increases the amount of work maintenance crews must perform to ready an aircraft for its next sortie. This additional maintenance time must be added to the additional flight time from the longer sorties to determine the average time between sorties (see Appendix B). In other words, as mission length increases, the average sortie-generation rate decreases, because an aircraft's second sortie cannot be launched until it returns from the longer first sortie and all required maintenance is performed.
- An additional constraint exists for "12-hour aircraft," aircraft such as F-15Es, F-117s, and A-10s that, for various reasons (e.g., survivability and system limitations), operate exclusively (or almost exclusively) during darkness or daylight. During Operation Desert Storm, all F-117 sorties and over 90 percent of F-15E sorties were flown at night to take advantage of advanced sensors while maximizing aircraft survivability, whereas the vast majority of A-10 sorties were flown during the day. The A-10s lacked sophisticated electronic sensors necessary to detect and attack targets effectively at night.

Sortie generation for aircraft with such restricted operating windows tends to stairstep as a function of the range to target. Initially, these aircraft might be able to fly two sorties per 12-hour period, but not a third sortie. As the distance to the target increases, they continue to be able to fly two sorties per day until the combination of flight time and ground maintenance time increases to the point where they can fly only one sortie in a 12-hour period. At this point, their sortie-generation capability abruptly falls by 50 percent. Figure 5.2 illustrates how the weapon-delivery capability of eight attack and multirole fighter wings, plus 60 bombers, varies as the range from the fighter bases to the target area increases.[1]

In addition to the 60 bombers, the strike force includes all of the USAF's combat-coded F-15Es and F-117s—124 and 47 aircraft, respectively—and approximately 50 percent of the USAF's active-duty inventory of F-16s.[2] We assume that all F-16 sorties are non-SEAD (suppression of enemy air defense) air-to-ground attack missions. To the extent that these multirole aircraft are used for air-

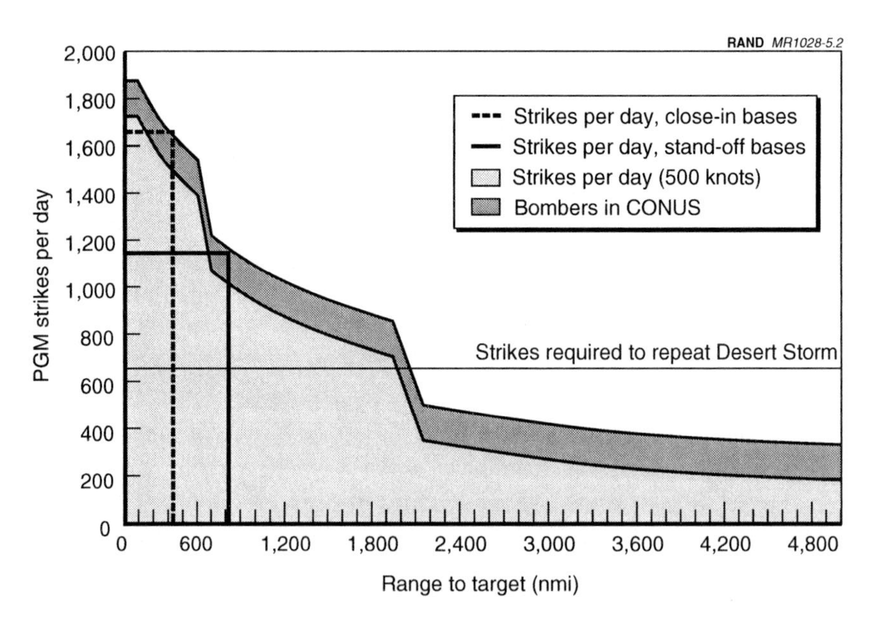

Figure 5.2—PGM Strikes As a Function of Range for a Postulated USAF Theater Attack Force

---

[1]For a more detailed description of the sortie-rate model used to generate Figure 5.2, see Appendix B.

[2]Aircraft numbers represent totals of primary aircraft authorized as of September 30, 1996. These are aircraft provided for the performance of operational missions. See *US Air Force Statistical Digest* (1996).

to-air or SEAD missions, our calculations overstate their capability to attack ground targets. Therefore, our calculations constitute a best-case analysis, with strike capability decreasing as the SAM and/or fighter threat increases. This strike force is probably at or near the maximum number of aircraft the USAF would commit to a single conflict.[3]

Figure 5.2 shows that, by devoting the lion's share of its resources to a single contingency, the USAF can easily repeat an air campaign of the size and intensity of Desert Storm from bases outside the range of the missile threat we have proposed.[4] In fact, the figure shows that the USAF could probably generate enough combat sorties from bases as far away as about 2,000 nmi from the target area. Beyond this range, round-trip missions will exceed 8 hours at 500 knots; as well, crews will probably have to be allowed time off between missions during a sustained campaign, since crew duty days—including mission planning, briefing, aircraft preflight, and debriefing—would approach 14 hours.

From the heavy dashed line in Figure 5.2, which shows the average number of strikes the 8-wing attack force could generate from the four close-in bases, and the heavy solid line, which illustrates the estimated strike potential of the same force operating from the more-distant bases, we can see that operating from more-distant bases cuts sortie generation and, therefore, weapon-delivery potential, by about 30 percent relative to closer bases.

From the outset of a conflict, the USAF would probably have to adopt a stand-off posture and accept the resulting reduction in combat power, since deploying to close-in bases and then falling back to bases outside missile range after absorbing an attack by ballistic or

---

[3]This level of commitment would leave the USAF and its Reserve components with a total of about 600 multirole fighters available to respond to other crises and support commitments to allies in other areas of the world (e.g., Korea). While these 600 aircraft would be a potent aerial strike force, they would not have anything close to the range or the weapon-delivery capability of the notional strike force described above.

[4] We use the same definition of a *strike* as presented in Cohen, GWAPS, 1993, Vol. V, p. 403: an attack by a single aircraft on a single target. The aircraft may use one or more weapons in its attack and, if it has enough weapons, may conduct multiple strikes in a single sortie. See the next section for a detailed description of how we arrived at the number of precision-guided-munition (PGM) strikes required to repeat a Desert Storm–size air campaign.

cruise missiles would be pointless. Adopting a go-in-close-and-fall-back-if-necessary strategy could result in the loss of a sizable fraction of USAF combat aircraft and personnel. Additionally, it may be difficult or impossible to move units out of close-in bases, because the continuing threat of attack would pose a grave threat to airlift operations, and most operating surfaces would be strewn with wreckage, posing a serious FOD hazard to surviving combat aircraft.

Figure 5.2 also shows that, over time, it may be possible for adversaries to force the USAF to operate from bases so far from the scene of a conflict that even if it commits the bulk of its forces, the USAF cannot sustain an air campaign of the scope and intensity of Desert Storm. This may not mean that the United States will lose a particular conflict, but the Desert Storm effort does at least represent a level of intensity that quickly defeated a moderately large and sophisticated adversary. Therefore, that effort represents a not-unreasonable yardstick for measuring the USAF's potential contribution in a major theater war (MTW). If enemy missile attacks could seriously degrade sortie generation—either by damaging close-in bases or pushing the USAF to more-distant operating locations—they could seriously impede attainment of important operational goals.

LONG-TERM STAND-OFF OPTIONS

If the USAF knew with a high degree of certainty that it could fight its next theater air campaign from permanent bases located on U.S. territory or the territory of a close ally such as the United Kingdom (UK), many of the problems associated with defenses designed to defeat the sorts of cruise and ballistic missiles we have described would be simplified or eliminated. This is especially true for fixed hardened shelters for aircraft and personnel: If the USAF knew where it was going to fight from, it could invest in protective facilities without fear that they would be rendered irrelevant by the need to respond to a crisis at some unexpected location.

Unfortunately, even with air refueling, current USAF combat aircraft probably could not sustain *intense combat operations* (which we define as one sortie per day per aircraft) over a distance of more than about 2,000 nmi; such missions simply take too long. A 2,000-nmi mission requires about 8 hours for an aircraft cruising at 500 knots to complete—an extremely long mission for fighter crews. Increasing the cruise speed of USAF attack aircraft would therefore increase the 8-hour mission radius. Figures 5.3 and 5.4 show the areas within reach of a high-spread force[5] (cruising at 500 and 1,000 knots, respectively) conducting 8-hour sorties from four "secure" bases: three on U.S. territory—Anderson AFB on Guam; Elmendorf AFB outside Anchorage, Alaska; and Homestead AFB near Miami, Florida—as well as secure and virtually permanent access to a facility such as RAF Lakenheath outside London, UK.[6]

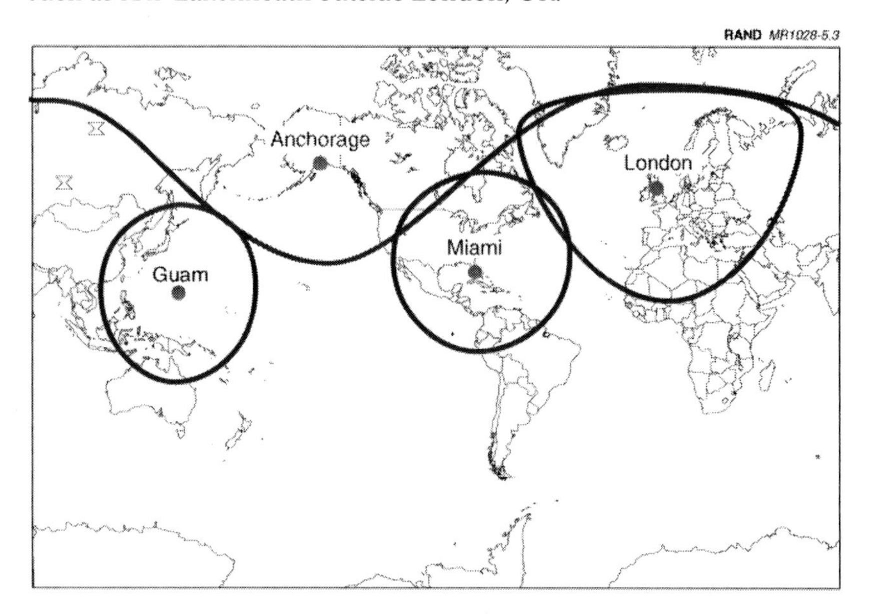

Figure 5.3—Areas That Current 500-Knot Fighter Forces Could Reach from Four "Secure" Bases

---

[5]See Appendix C for a more complete analysis of the high-speed-attack-aircraft concept.

[6]We picked four bases to illustrate the advantages of long-range aircraft. The actual number of bases for the force would have to be derived from an analysis of a host of operational, logistics, political, and cost factors. Whatever the final number, efficiency concerns will keep the number under 10. Although suitable base locations would require further study, the four bases chosen for this analysis are about as far apart from each other as possible, giving the most global coverage. An additional base at Diego Garcia would increase global coverage slightly by bringing the southern tip of Africa within range, and would provide additional operational flexibility and multiple ingress and recovery options for missions against targets in much of Asia. Expanding basing options to all NATO countries might improve the coverage in parts of Africa and Asia.

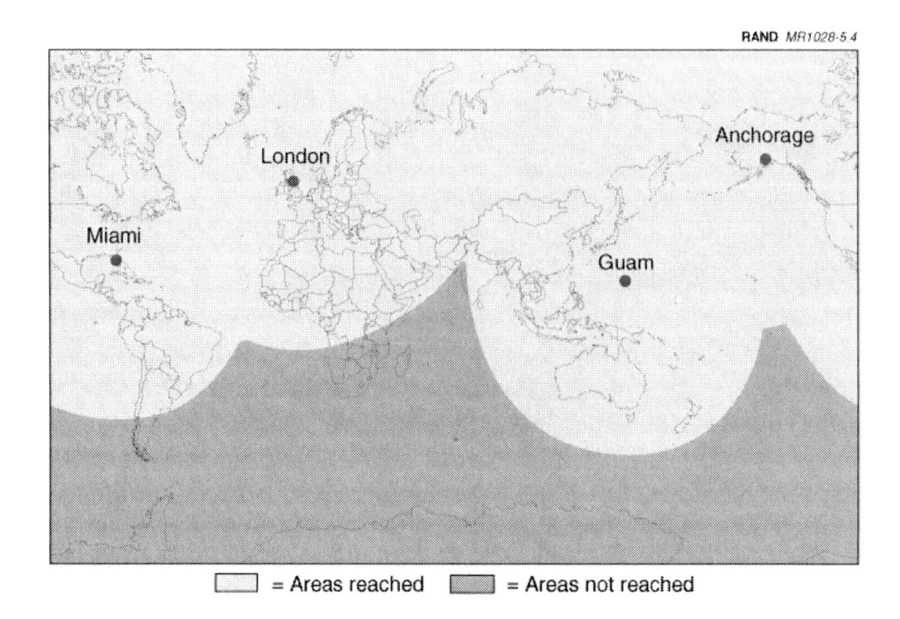

Figure 5.4—Areas of the World a Force of 1,000-Knot Aircraft Could Reach from Four Secure Bases

Obviously, concentrating valuable aircraft at a small number of bases would create very lucrative targets. Although few countries today possess conventional systems that could attack installations in the U.S. and UK locations we suggest, more may acquire such capabilities in the future. However, three things seem clear:

- First, long-range strike assets will almost certainly remain more technologically challenging and expensive—and therefore less numerous—than shorter-range systems. Therefore, bases distant from likely theaters of action should remain under a quantitatively lesser threat than those close in.[7]
- Second, the escalation implications of attacking bases on U.S. territory—and, to a certain extent, on British soil—could play a

[7]So much so that we believe the smaller numerical magnitude of the threat would *qualitatively* reduce the risk to USAF assets at the distant bases. As Lenin said, "Quantity can have a quality all its own."

role in deterring opponents from undertaking such strikes in the first place.

- Third, by operating forces from a small number of assured, known facilities, it may be practical to undertake a number of passive and active protection measures, such as hardening and anti-missile defenses, that would be infeasible in today's world of dependence on uncertain forward basing.

Another proposal that has been discussed to reduce the USAF's dependence on short-range aircraft is the "arsenal plane,"[8] which could launch large numbers of accurate, long-range cruise missiles that could effectively operate over very long ranges. But this is only one way to achieve rapid and robust global conventional attack capability.

We recommend that a concept centered around fast, long-range aircraft be considered. Not only does such a concept address the need to protect USAF assets from the ever-expanding range of enemy threat systems, it also promises to greatly ease the kinds of access burdens that have from time to time troubled U.S. power-projection operations.[9] Some of the technical issues surrounding such an aircraft are discussed in Appendix C. Furthermore, we recommend that a more detailed study be undertaken to compare this idea with other options (such as 747 "arsenal planes") for the USAF to determine which concept is most technically sound, operationally effective, and fiscally viable.

---

[8]The Boeing Company, *747 Air-Launched Cruise Missile System Concept,* Seattle, Wash., April 1974.

[9]Saudi Arabia's reported refusal to support U.S. punitive operations against Iraq in early 1998 is only the most recent public example of these bedevilments.

Chapter Six

CONCLUSIONS

This report has shown how competent and committed adversaries could take advantage of the way in which the USAF plans to conduct theater air operations. They could do so by combining several well-known and widely available technologies, such as UAVs, GPS, sub-munition warheads, and ballistic missiles. Armed with accurate area weapons capable of inflicting substantial damage on soft-skinned targets, such as parked aircraft, tents, radars, and personnel, such an adversary could severely disrupt the USAF's ability to conduct combat operations from theater airbases.

Over the short term, the USAF could respond to this threat by deploying its combat forces with extensive active and passive defenses. However, these measures may be too manpower-intensive, heavy, or expensive to protect all USAF assets required to support combat operations. In addition, some very valuable USAF aircraft, such as airlifters, bombers, and intelligence, surveillance, and reconnaissance (ISR) platforms, will be very difficult if not impossible to protect passively because of their size. Another alternative is to operate from dispersed locations or to disperse USAF aircraft more widely at existing facilities. These concepts could work to reduce the effectiveness of a missile attack against USAF assets but have significant drawbacks of their own, such as decreased USAF sortie-generation efficiency, increased logistics costs and complexity, the need to accurately predict the scene of future conflicts, and failure to address the threat posed to tent cities.

Another possible response to long-term increases in adversary missile capability is to shift away from an operational concept that requires large numbers of USAF fighters and their support personnel to deploy to a combat theater. Instead, the USAF could rely on a fleet of long-range aircraft operating from permanent bases to project USAF combat power. Appendix C describes a concept for fast, high-flying, stealthy bombers that could operate in this way.

The options we have discussed here for conducting attacks against USAF theater airbases, and the possible USAF defensive responses, are almost certainly incomplete. There are probably other ways to attack the bases, and other options (or combinations of options) for dealing with the attack methods we have presented. However, it is not important that our descriptions of various attack options represent absolutely accurate predictions of future adversary attack plans or capabilities. What is important is that USAF planners recognize that the success enjoyed by the USAF during Desert Storm provided a powerful incentive for any nation considering military action that could involve conflict with the United States to devise an effective plan for dealing with USAF land-based air combat power.

We have shown how such an adversary could go about constructing a simple, robust, effective, and relatively affordable capability to disrupt USAF theater air operations. The options presented here represent only the beginning of a long process of systematically laying out and analyzing the various options available to the USAF for decreasing its vulnerability to such attacks.

Appendix A

DAMAGE CALCULATION FOR PARKED AIRCRAFT

This appendix describes the method we used to calculate the number of ballistic and cruise missiles required to destroy aircraft parked on airbase ramps. The method allowed us to look at a significant number of cases and facilitated a simple and fast spreadsheet analysis of the expected number of missiles that would be required. We believe the answers are "ballpark estimates" that show the magnitude of kills that can be expected. Further refinement of the method is required. This analysis was conducted using publicly available unclassified information of likely U.S. deployment bases in Southwest Asia (SWA).[1] Technical data on ballistic- and cruise-missile systems also was obtained from unclassified sources.[2]

The calculations require a description of the parking ramps at each airbase and the accuracy and lethality of the missiles used. In this appendix, we first describe the input data and simplifying assumptions used, then describe the calculations for unitary warheads. Next, we discuss the method used to introduce submunition warheads into the analysis. Finally, we present sample results.

---

[1]The authors chose the Southwest Asia deployment bases having superior facilities and based on historical precedent.

[2]Duncan Lennox, ed., *Jane's Strategic Weapon Systems*, Coulsdon, Surrey, England: Jane's Information Group, Issue 24, May 1997.

DATA AND ASSUMPTIONS

Parking Areas

We obtained data on potential parking areas on an airbase from *DOD Flight Information Publication: High and Low Altitude Europe, North Africa, and Middle East.*[3] This information is also available as a computer database called the Digital Aeronautical Flight Information File (DAFIF). These documents are available to the general public at the unclassified level.

Figure A.1 presents a sample diagram of the airbase at Dhahran. Using such scale drawings, we obtained the dimensions of each of the potential parking areas on the base. The calculation method we used assumes that the parking areas are rectangular, which is typical of many bases throughout the world.

From the diagram of Dhahran Airbase, we chose the following rectangular parking areas for attack (all dimensions are in feet):

parking area 1: 9,000 × 900

parking area 2: 4,200 × 600

parking area 3: 900 × 900

parking area 4: 1,200 × 900

parking area 5: 2,100 × 700

Missile Characteristics

The calculations used the lethal radius, l, circular error probable, *CEP*, and range of potential systems, R. We made simplifications by assuming that the weapons are "cookie cutters": They have 100-percent effectiveness inside the lethal radius and 0-percent

[3]National Imagery and Mapping Agency, Vol. 5, St. Louis, Missouri, February 27, 1997.

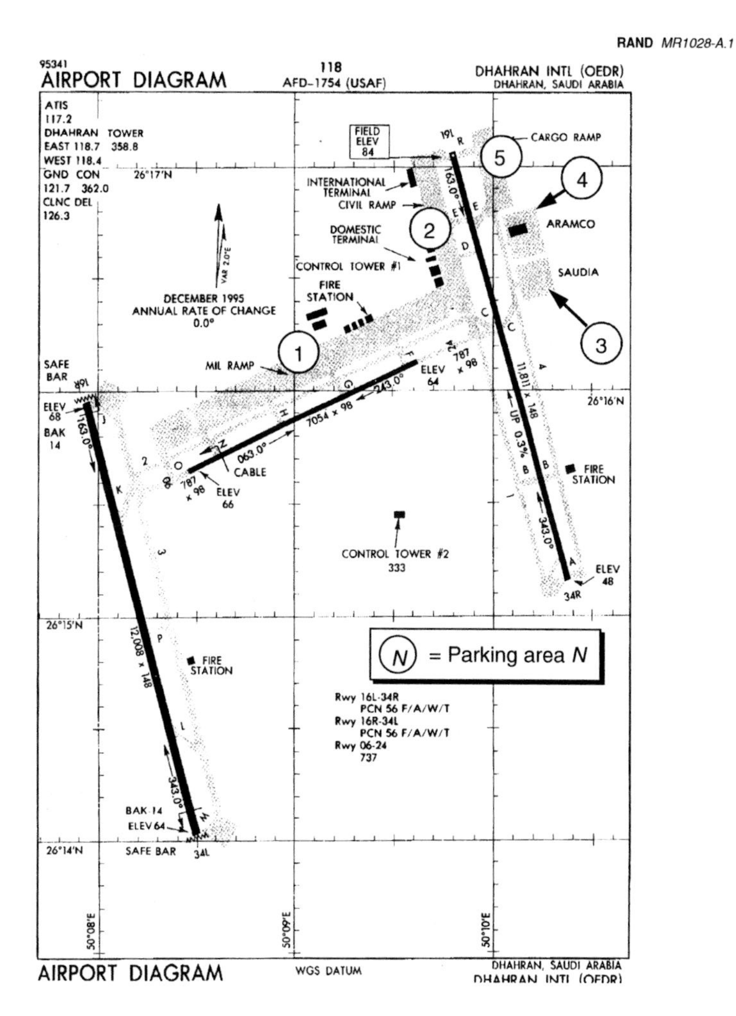

SOURCE: National Imagery and Mapping Agency, *DoD Flight Information Publication: High and Low Altitude Europe, North Africa, and Middle East,* Vol. 5, St. Louis, Missouri, February 27, 1997.

Figure A.1—Diagram of Dhahran Airbase

effectiveness outside, as in the circles in Figure 2.3.[4] The difference in downrange and cross-range errors was ignored, and the CEP was used as the measure of accuracy.

Weaponeering Calculations

Each parking area was assumed to be composed of a series of squares, or *blocks,* each with the dimension of the minimum length of the parking ramp. For example, the 4,200 × 600 foot parking ramp at Dhahran, parking area 2 in Figure A.1, was assumed to be composed of seven 600 × 600 foot blocks. For the purpose of obtaining rough estimates and automated calculations, we placed the aimpoint of each weapon at the center of its respective calculation block (as shown in Figure A.2).

The Reason for the Two-Calculation Regimes

The most important factor in this analysis is the size of the weapon's lethal area relative to that of the target calculation block. Two possible relationships lead to different calculation methods, or regimes: (1) the weapon's lethal area is smaller than or on the same order as the area of the calculation block or (2) the weapon's lethal area is larger than the area of the calculation block. These assmptions result in ballpark estimates, but they allowed us to do quick calculations. In some places, combinations of *l, CEP,* etc., can give less-accurate results. We performed no analyses of the extent of the accuracy.

In the first case, one weapon will not be enough to cover the calculation block, no matter how accurately it can be delivered. In the second, one accurate weapon may do the job; additional weapons are required only when the first does not have the necessary accuracy. We considered the two regimes separately. Later in this appendix,

---

[4]In the case of a submunition warhead, the "cookie-cutter"-weapon-effect assumption means that at least enough submunitions are distributed throughout the pattern to have a 100-percent probability of kill for the target being considered and no submunitions fall outside the pattern. This method should provide fairly accurate results that err slightly on the conservative side in kill potential against large-area targets.

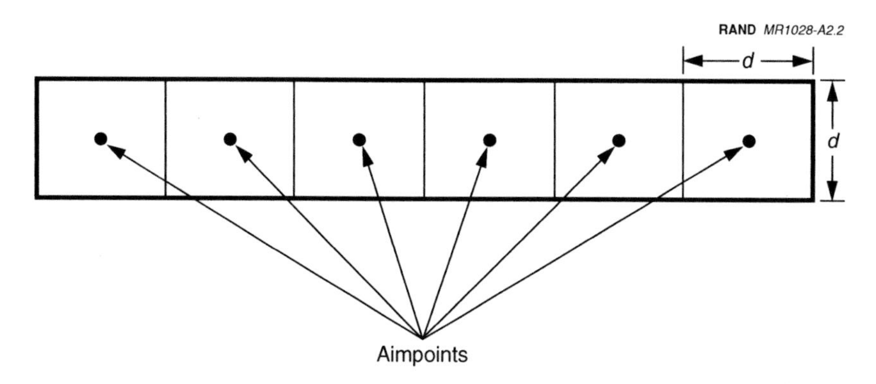

Figure A.2—Layout of Calculation Blocks

we show how analyzing these two regimes separately enabled us to make assumptions that greatly simplify the calculations so that they lend themselves to fast (and easily automated) calculation methods.

We define the first regime to be a case where the lethal radius of the weapon is less than or equal to half of the minimum dimension of the ramp, *d*. The second is the case where the lethal radius is greater than half of the minimum dimension, *d*. For both regimes, we assumed a "salvo" attack on the ramp: One weapon was aimed at the center of each calculation block for every salvo, and the proportion of the block covered by blast and fragments was computed. The number of salvos required to achieve a given probability of kill was then calculated.

Regime 1: Weapon's Effect Small Relative to Block ($l \leq d/2$)

Edge Effects. Before we get into a discussion of the calculation, we address weapon edge effects. In Figure A.3, the area of the white square and its dark-shaded border represents one of the calculation blocks discussed above (i.e., one of the seven 600 × 600 foot blocks on the 4,200 × 600 foot ramp at Dhahran). The white block represents the area in which the impact point of a weapon could lie and within which the weapon effect would be entirely contained. Impact points within the dark-shaded border area (Area 2) and the light-shaded border area (Area 3) surrounding it result in only partial coverage of the calculation block by the weapon.

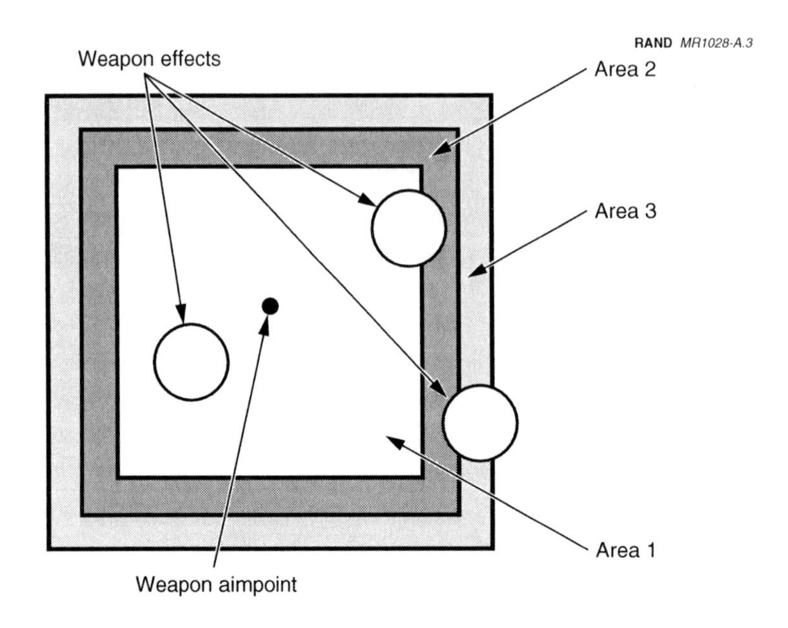

Figure A.3—Diagram for Edge-Effects Analysis

Although a weapon has a slightly higher probability of landing in Area 2 than in Area 3, we neglect this difference in order to simplify the calculations: We used the entire weapon effect when the impact point lies in Area 2, but no weapon effect when the impact is in Area 3.

Diagrams and Terminology. For the following discussion of calculations, we show how parking ramps were divided and grouped. Earlier, Figure A.2 showed a $6d \times d$ aircraft parking ramp divided into 6 calculation blocks. The expected area covered by one salvo must be calculated for three block types (interior, transition, and edge). In Figure A.4, the number of blocks, n, is now 10. We allowed the CEP inherent in all weapons to distribute multiple weapons within each block. These errors are assumed to be circular and to be normally distributed about the aimpoint, with the magnitude of the dispersion determined by the CEP. The variable "NormDist" in the equations

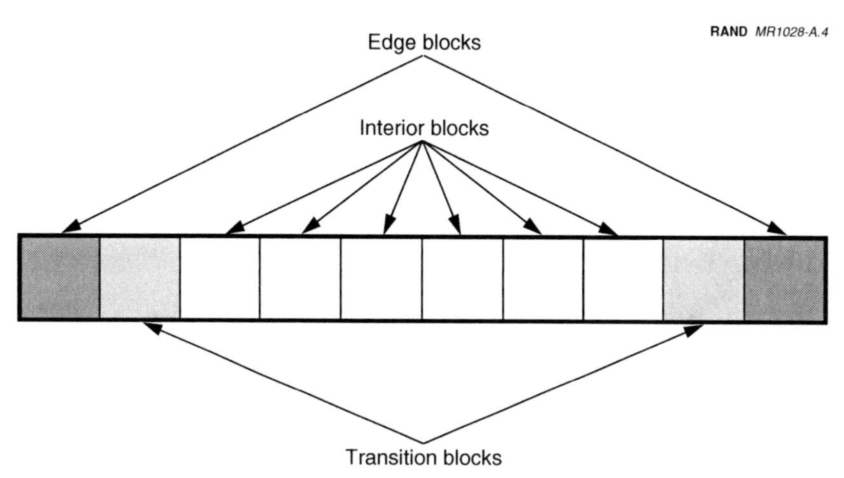

Figure A.4—Block Terminology

that follow indicates the probability of occurrence using a circular normal distribution.[5]

Calculating Area of First Salvo. A *salvo* is defined as one weapon fired at the center of each block at the same time (as in Figure A.2 above).

For each weapon in the first salvo, there are three impact possibilities:[6]

- The weapon may land in the block at which it was aimed.

---

[5]We chose this method of aimpoint selection, since the CEPs are large (at least on the same order) relative to the lethal radius for the case being examined in this analysis. We are simply aiming the weapons on the ramp. This simple aimpoint-choice method fails when the CEP is small, since we no longer get coverage of the entire ramp. Of course, this is not a problem in the real world: If we were deploying very accurate weapons, we would simply choose more than one aimpoint in a given calculation square. This assumption merely allows us to "automate" the aimpoint selection for these calculations (such "automation" uses the naturally occurring probabilistic spread of the weapons to cover large ramps, so that the aimpoint does not have to be selected for each weapon).

[6]We do not consider the possibility of weapon overlap for weapons in the same salvo. Weapon overlap is considered later, when the number of salvos required to achieve the required coverage is calculated.

- It may land in a different block.
- The weapon may land completely off the ramp.

To simplify the calculations, we first assumed that each block was a circle of equivalent area (we "circled the block"), with the aimpoint representing the area's center, as shown in Block 1 of Figure A.5:

$$r = \sqrt{\frac{d^2}{\pi}} = \frac{d}{\sqrt{\pi}} \tag{A.1}$$

Such "circling" should affect the calculation's validity minimally.

Next, we consider the probability of the weapon's landing in the block at which it was aimed (Block 1 in Figure A.5). For our circularized blocks and a circular weapon with circular normal-distribution impact errors, this probability is

$$P_{\text{Block 1}} = 2(\text{NormDist}_r - \text{NormDist}_0) \qquad \text{(A.2)}^7$$

The expected area, A, covered in Block 1 by this weapon is

$$A_{\text{Block 1}} = P_{\text{Block 1}} \pi l^2 \tag{A.3}$$

where l is the lethal radius of the weapon.

Next, we consider the possibility of a weapon's landing in a different block from the one at which it was aimed. Again, we assume Block 1 to be a circle with radius r as shown in Figure A.5. A circle with radius $3r$ encompasses both Block 1 and Block 2 (transition block in Figure A.4). An estimate of the probability that the weapon aimed at the center of Block 1 strikes Block 2 can be obtained by first calculating the probability that the weapon will strike in the toroid between a circle with radius r and a circle with radius $3r$. Then, since we are assuming a circular distribution, the probability that the weapon will land in Block 2 is simply the ratio of the area of Block 2 to that of the toroid:

---

[7]NormDist is the cumulative probability distribution for the value in the subscript.

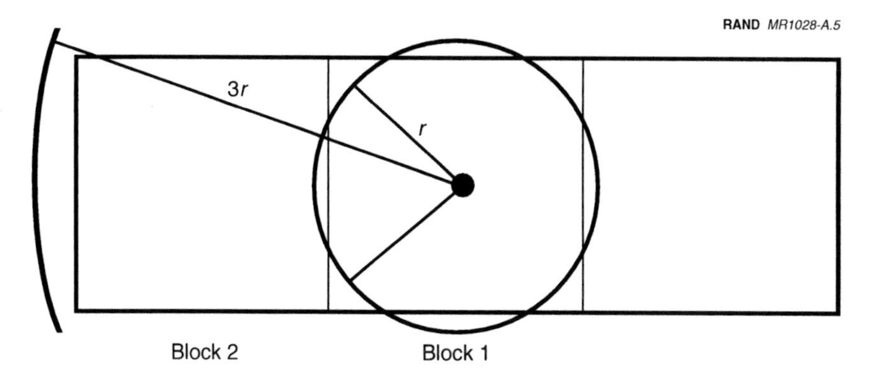

Figure A.5—Circular Patterns Superimposed on Calculation Blocks for Same-Block and Different-Block Calculations

$$P_{\text{Block 2}} = \left(\text{NormDist}_{3r} - \text{NormDist}_{r}\right) \frac{A_{\text{Block 2}}}{A_{3r} - A_{r}} \tag{A.4}$$

Therefore,

$$P_{\text{Block 2}} = \left(\text{NormDist}_{3r} - \text{NormDist}_{r}\right) \frac{\pi r^2}{\pi\left[\left(3r\right)^2 - r^2\right]} \tag{A.5}$$

$$P_{\text{Block 2}} = \frac{1}{8}\left(\text{NormDist}_{3r} - \text{NormDist}_{r}\right) \tag{A.6}$$

The expected area covered in a block by a weapon that was aimed at an adjacent block is simply the probability that it will strike the block multiplied by the lethal area of the weapon (remember, we are neglecting edge effects). That is,

$$A_{\text{Block 2}} = P_{\text{Block 2}} \pi l^2 \tag{A.7}$$

The final case to be considered here is the expected area covered in a block by a weapon that was aimed at the center point of a block two blocks away, presented graphically in Figure A.6 (the edge block in Figure A.4).

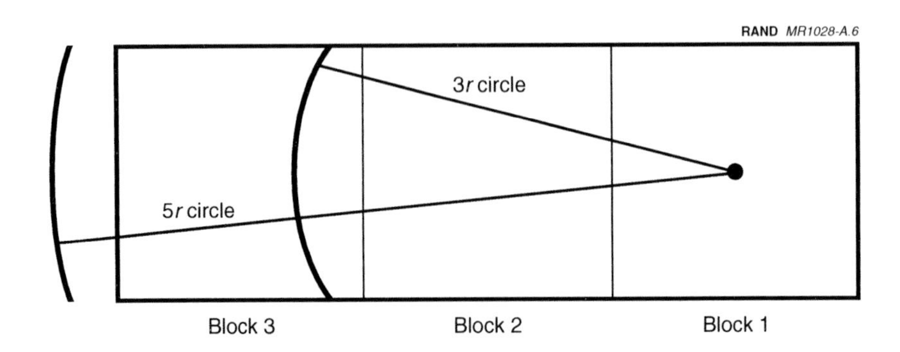

Figure A.6—Area Covered by Weapon Aimed at Neighboring Blocks

An estimate of the probability that the weapon aimed at the center of Block 1 strikes Block 3 can be derived by first determining the probability that the shot lands in the toroid between the circles with radii of $3r$ and $5r$, then multiplying that value by the area of Block 3 divided by the area contained in the toroid. Thus,

$$P_{\text{Block 3}} = \left(\text{NormDist}_{5r} - \text{NormDist}_{3r}\right)\frac{A_{\text{Block 3}}}{A_{5r} - A_{3r}} \quad \text{(A.8)}$$

$$P_{\text{Block 3}} = \left(\text{NormDist}_{5r} - \text{NormDist}_{3r}\right)\frac{\pi r^2}{\pi\left(5r^2 - 3r^2\right)} \quad \text{(A.9)}$$

$$P_{\text{Block 3}} = \frac{1}{16}\left(\text{NormDist}_{5r} - \text{NormDist}_{3r}\right) \quad \text{(A.10)}$$

The expected area in a block affected by a shot that was aimed at the center of a block located two blocks away is again calculated by multiplying the probability by the lethal area of the weapon:

$$A_{\text{Block 3}} = P_{\text{Block 3}}\pi l^2 \quad \text{(A.11)}$$

Next, we determine the expected coverage by a salvo of weapons fired at the parking ramp. The number of blocks is

$$n = \frac{x}{d} \tag{A.12}$$

where x is the length of the long axis of the parking ramp and d is the length of the short axis. Figure A.4 identifies the terminology used in the discussion below. In this example, $n = 10$. The expected area covered by one salvo must be calculated for three block types (interior, transition, and edge) individually. The expected area covered in an interior block is

$$A_{\text{interior}} = A_{\text{Block 1}} + 2A_{\text{Block 2}} + 2A_{\text{Block 3}} \tag{A.13}$$

where $A_{\text{Block 1}}$, $A_{\text{Block 2}}$, and $A_{\text{Block 3}}$ refer to the calculations above. Recall that $A_{\text{Block 1}}$ is the expected area of the block in question being covered by a weapon that was aimed at that block. $A_{\text{Block 2}}$ is the expected area of the block in question being covered by a weapon that was aimed at an adjacent block, and $A_{\text{Block 3}}$ is the expected area of the block in question being covered by a weapon that was aimed at a block on the other side of an adjacent block. The coefficients in the equation above account for the number of each of these types of blocks. We could extend this calculation to account for weapons that were aimed at blocks farther away, but to do so would add needless complexity to our calculations, given that the probabilities quickly drop off by the very nature of the normal distribution and that the ratio of block area to toroid area is reduced with each successive block away. Therefore, the probability that a weapon with an aimpoint more than two blocks away affects targets in a calculation block is small.

The expected area covered in a transition block and in an edge block follows by applying this methodology in the context of Figure A.4:

$$A_{\text{transition}} = A_{\text{Block 1}} + 2A_{\text{Block 2}} + A_{\text{Block 3}} \tag{A.14}$$

$$A_{\text{edge}} = A_{\text{Block 1}} + A_{\text{Block 2}} + A_{\text{Block 3}} \tag{A.15}$$

Therefore, the total expected area covered by a salvo of N weapons (one aimed at the center point of each of the n blocks) is

$$A_{\text{total}} = (n-4)\,A_{\text{interior}} + 2A_{\text{transition}} + 2A_{\text{edge}} \tag{A.16}$$

Making the above substitutions and simplifying the equation, we are left with

$$A_{\text{total}} = n\left(A_{\text{Block 1}} + 2A_{\text{Block 2}} + 2A_{\text{Block 3}}\right) - 2A_{\text{Block 2}} - 4A_{\text{Block 3}} \tag{A.17}$$

This calculation accounts for N weapons aimed at the center point of each of n blocks, not for the possibility of overlapping weapons; therefore, the area affected is slightly high. However, expecting no overlap is a fairly good assumption when the lethal radius of the weapon, l, is small relative to the size of the blocks. With this assumption, the expected fraction, F, of ramp covered by one salvo is

$$F_{(1\text{ salvo})} = \frac{A_{\text{total}}}{xd} \tag{A.18}$$

Calculation of Salvos Required for Specified Level of Damage. Now we can calculate the number of salvos required to achieve a specified level of damage on the ramp. For these calculations, we can consider the possibility of overlap. The total coverage that can be expected from N salvos is defined as

$$F_{(N\text{ salvos})} = 1 - \left[1 - F_{(1\text{ salvo})}\right]^N \tag{A.19}$$

Therefore, the number of salvos, N, required to achieve a specified coverage, $F_{\text{coverage req'd}}$, is

$$N = \frac{\log\left(1 - F_{\text{coverage req'd}}\right)}{\log\left[1 - F_{(1\text{ salvo})}\right]} \tag{A.20}$$

Regime 2: Weapon's Effect Large Relative to Block ($l > d/2$)

Now we turn our attention to the case where the lethal radius is larger than $d/2$. This regime is analyzed differently, since one weapon could cover more than one block. We again analyze the problem as a number of blocks with dimensions of the minimum size of the ramp. Again, we assume that one shot is aimed at the center of each of the blocks, and we determine the expected area covered. In this case, one shot can entirely cover one or more blocks. The most important simplifying assumption of this analysis is that a block is considered to be covered if the center point falls within the "cookie cutter" effect of the weapon, shown in Figure A.7. Again, the distribution of weapon impact points was also assumed to be circular following a normal distribution, as in Regime 1.

We first consider the possibility that a weapon covers the block at which it was aimed. The probability that the center of the block is covered (and, according to our assumptions, the block itself) is the probability that the weapon lands within one lethal radius of its aimpoint (the center of the block). The probability of coverage is

$$P_{\text{Block 1}} = 2\left(\text{NormDist}_l - \text{NormDist}_0\right) \tag{A.21}$$

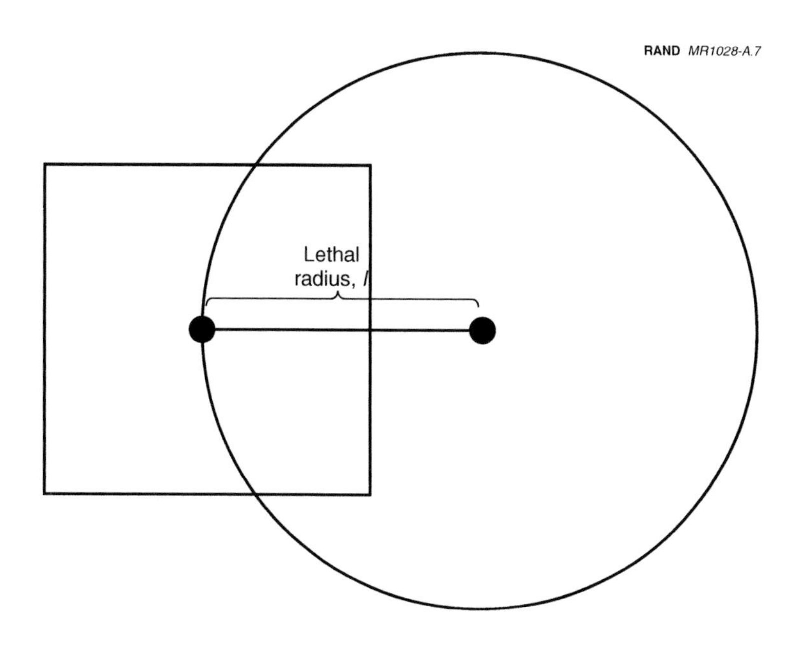

Figure A.7—Calculation Based on Coverage of Center Point of Block

where l is the lethal radius of the weapon. The expected area covered is

$$A_{\text{Block 1}} = P_{\text{Block 1}} d^2 \tag{A.22}$$

Now, turning our attention to adjacent blocks, we consider the probability that this weapon also covers neighboring blocks. Again, we consider a block to be 100-percent covered if its center is contained within the lethal radius of the shot. In this case, we are looking for the probability that Block 2 will get covered by a weapon aimed at Block 1. Recall that when considering the probability of a weapon's covering a point, if the weapon lands within the distance l from that point, the point is covered. So, since we want to analyze the probability that a weapon will cover the center point of Block 2, we draw a circle of radius l around that point and recognize that a weapon whose impact point is within that circle will cover the center of Block 2. Given that, the question becomes, "What is the probability that a weapon will impact within the circle?" That problem is graphically presented in Figure A.8 by the normal distribution superimposed on the ramp diagram. For the random variable angle ϕ chosen, we can now determine the probability of the impact point of the weapon aimed at the center of Block 1 landing within the circle of radius l centered at Block 2. We do so by computing the area under the curve and between a and b. Now, all values of ϕ are equally likely, but, as can be seen in Figure A.8, different values of ϕ produce different probabilities that the weapon will strike the center of Block 2. Therefore, to determine the probability of a shot aimed at Block 1 landing within the circle and thus covering Block 2, we must determine the probability of a shot's landing between a and b for each angle of ϕ and then average across all ϕ.

Figure A.9 presents the method used to determine the values of a and b for each angle of ϕ. The value of d is known, since it is the distance between the aimpoint and the center of the block under consideration, and is the same value d used throughout the preceding 22 calculations, since the distance between the center points of two blocks is equal to the dimension of each of the blocks.

In this figure, the two congruent triangles from which a and b are measured are along the diagonal line that is inclined at angle ϕ to the

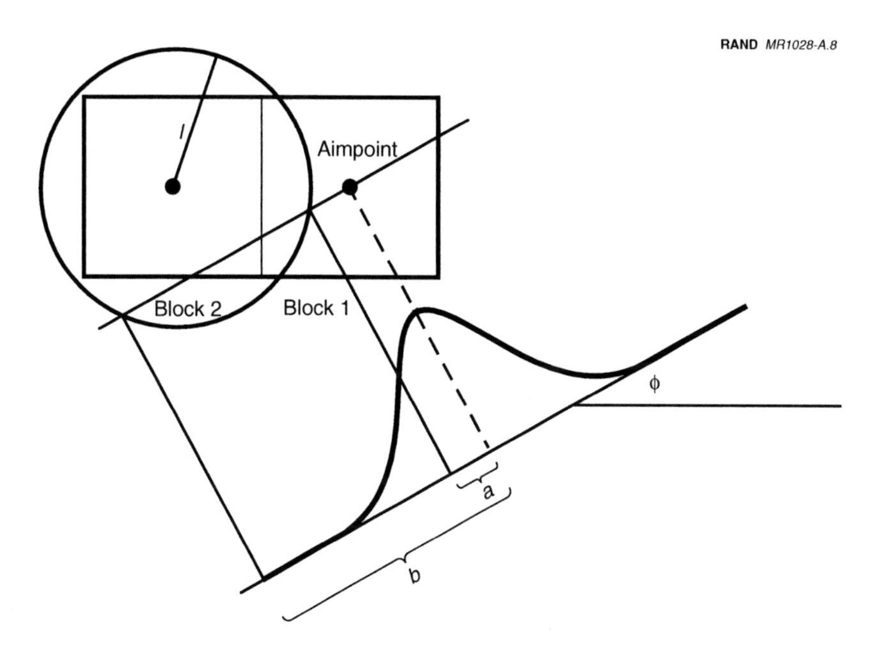

Figure A.8—Diagram for Calculation of Neighboring-Block Coverage

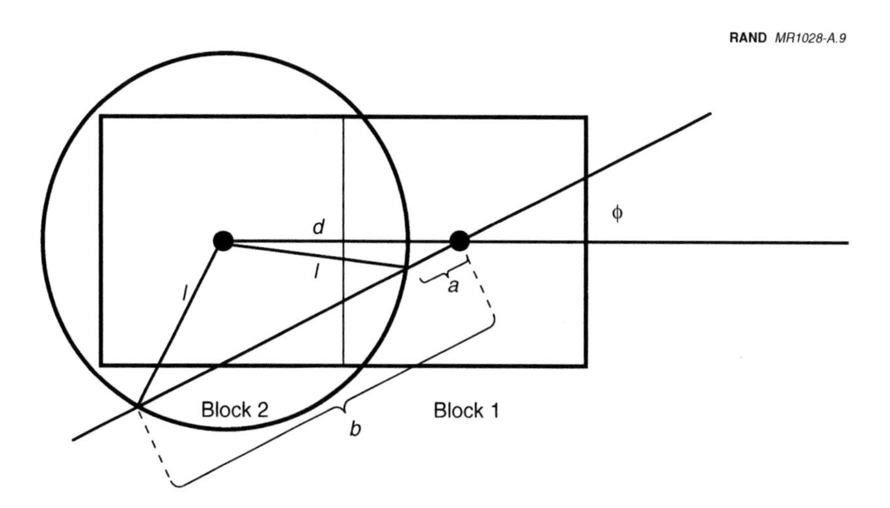

Figure A.9—Geometry of Neighboring-Block Calculation

horizontal; therefore, the values of a and b can be derived using the law of cosines:

$$l^2 = a^2 + d^2 - 2ad \cos \phi \quad \text{(A.23)}$$

and

$$l^2 = b^2 + d^2 - 2bd \cos \phi \quad \text{(A.24)}$$

Noting that the above equations are in quadratic form, we obtain the solution by applying the quadratic equation:

$$a,b = d \cos \phi \pm \sqrt{d^2 \cos^2 \phi - d^2 + l^2} \quad \text{(A.25)}$$

The same method can be applied to determine the probability of covering a block that is located two or more blocks away from the aimpoints by replacing d with $2d$, $3d$, . . ., nd in the above equations. Once a and b are determined, the probability of a shot's landing between these two values for a given ϕ is

$$P_{\text{Block 2}} = \text{NormDist}_a - \text{NormDist}_b \quad \text{(A.26)}$$

Since we are assuming that all values of ϕ are equally likely as a result of the circular error distribution, the probability of a shot aimed at the center of Block 1 landing within a circle of radius l centered at Block 2 is the average of $P_{\text{Block 2}}$, calculated above, across all values of ϕ. $P_{\text{Block 3}}$ was determined using the same method and replacing d with $2d$.

As in Regime 1, the next step is to analyze the effectiveness of a salvo attack. We assume that a salvo of N weapons is fired at the ramp of n blocks with an aimpoint corresponding to the center of each block. For simplicity, the expected coverage of Block 1 is calculated as the expected coverage by the weapon that was aimed at Block 1 and the expected coverage from weapons that were aimed at each of the four most-proximal blocks. Unlike the analysis of Regime 1, the weapons have large coverage areas relative to the size of the blocks, and the possibility of overlap is high, i.e., two weapons aimed at two adjacent blocks could both cover both blocks. Again referring to Figure A.4 for the block terminology and accepting that we are limiting the calculation, for simplicity, to weapons aimed at a particular block and those that are aimed at blocks no more than $2d$ away, we get the following probability of coverage for each interior block:

$$P_{\text{interior}} = 1 - \left(1 - P_{\text{Block 1}}\right)\left(1 - P_{\text{Block 2}}\right)^2 \left(1 - P_{\text{Block 3}}\right)^2 \qquad \text{(A.27)}$$

where $P_{\text{Blocks 1, 2, 3}}$ are the probabilities calculated earlier in this discussion of Regime 2 calculations. The second and third terms in this equation are squared to account for the two blocks on each side of the block in question.

Likewise, the probability of coverage for the transition and edge block is

$$P_{\text{transition}} = 1 - \left(1 - P_{\text{Block 1}}\right)\left(1 - P_{\text{Block 2}}\right)^2 \left(1 - P_{\text{Block 3}}\right) \qquad \text{(A.28)}$$

and

$$P_{\text{edge}} = 1 - \left(1 - P_{\text{Block 1}}\right)\left(1 - P_{\text{Block 2}}\right)\left(1 - P_{\text{Block 3}}\right) \qquad \text{(A.29)}$$

Therefore the total expected area covered in the entire parking ramp for one salvo is

$$A_{\text{total}} = d^2[(n-4)P_{\text{interior}} + 2P_{\text{transition}} + 2P_{\text{edge}}] \qquad \text{(A.30)}$$

where the term d^2 is the area of one block. Given that the area of the entire ramp, A_{ramp}, is nd^2, the total expected coverage of the ramp becomes

$$F_{(\text{1 salvo})} = \frac{(n-4)P_{\text{interior}} + 2P_{\text{transition}} + 2P_{\text{edge}}}{n} \qquad \text{(A.31)}$$

Using the same procedure as that used in Regime 1 to determine the number of salvos, N, required to achieve a specified level of coverage, $F_{\text{coverage req'd}}$, we obtain

$$N = \frac{\log\left(1 - F_{\text{coverage req'd}}\right)}{\log\left[1 - F_{(1\ \text{salvo})}\right]} \tag{A.32}$$

Submunition Warheads

Since large overpressures[8] are not required in order to cause significant damage against soft targets such as aircraft in the open, submunition warheads are an ideal weapon. Submunition technology provides the capability to significantly damage aircraft over wide areas with relatively small warheads.

The measure of effectiveness of a warhead is the lethal radius within which damage or kill occurs. The lethal radius of a submunition warhead depends on the lethal radius of the individual submunitions and the efficiency of the submunition dispersal, or the ability of the weapon to distribute the individual bomblets uniformly over an area. The best choice for dispersal pattern varies markedly for different target types, e.g., troops in foxholes and aircraft in the open.

This analysis postulates a warhead designed for efficient aircraft attack. The warhead is assumed to disperse submunitions uniformly at large separation distances, since an aircraft is a large target and damage to any part can make it unflyable. The warheads are assumed to consist of a number of 1-pound fragmenting bomblets and the additional mechanics to contain and dispense the submunitions. We assume bomblets constitute 75 percent of warhead weight and that mechanical components and structure make up the remaining 25 percent.

Sample Results

This section presents some results based on the method described above. Table A.1 shows the number of ballistic and cruise missiles of each type required to attack each of the ramps at four Southwest Asia

---

[8] *Overpressure* is the higher-than-ambient pressure that occurs for a short time in the vicinity of the explosion.

Table A.1

Number of Ballistic and Cruise Missiles of a Given Type Required to Attack Each Ramp at Four Southwest Bases

		Scud C	M-9 Unit	M-9 Sub	NODONG	M-18 Unit	M-18 Sub	CSS-5	CM-1	CM-2
	Reliability	0.7	0.95	0.9	0.7	0.95	0.9	0.95	0.5	0.5
	CEP (ft)	2,394.4	656	656	6,068	656	656	656	328	328
	Lethal radius (ft)	241	206	862	260	206	862	260	232	141
	Fraction of kills required	0.9	0.9	0.9	0.9	0.9	0.9	0.9	0.9	0.9
	Cost ($M)	0.2	0.9	1	0.3	1.9	2	3	0.3	0.2
Length (ft)	Width (ft)									
Dhahran					Number of Missiles Required					
9,000	900	330	170	12	710	170	12	100	110	300
4,200	600	154	63	5	336	63	5	42	35	98
900	900	66	23	2	148	23	2	14	11	31
1,200	900	69	28	2	152	28	2	17	15	41
2,100	700	87	39	3	192	39	3	24	21	57
Doha										
2,100	700	87	39	3	192	39	3	24	21	57
600	300	41	14	2	93	14	2	10	6	12
600	600	44	15	2	98	15	2	9	6	17
1,800	900	86	38	3	186	38	3	24	22	62
Riyadh										
5,600	800	210	98	7	455	98	7	63	56	168
3,300	300	143	55	5	315	55	5	37	26	44
8,400	700	300	132	9	660	132	9	84	72	228
Al Kharj										
6,400	1,600	276	184	7	564	184	7	116	136	376
5,000	1,000	200	100	7	420	100	7	60	65	185

bases.[9] Every number in the matrix is independent of all others and represents the total number of a weapon type to cover a given ramp exclusively, with no other weapon types being employed against the ramp.

A simple linear program can then be constructed to design an optimal attack strategy within specific constraints. For example, a least-cost solution could be achieved with the constraint that each airbase be attacked by both ballistic and cruise missiles to complicate defenses.

---

[9]These calculations assume a lethal radius of 20 feet for a 1-pound submunition.

Appendix B

SORTIE-RATE MODEL

This appendix describes the method used to calculate aircraft sortie rates. Based on earlier unpublished work by former RAND colleague Lawrence Hollett,[1] the method was chosen to facilitate a simple and fast spreadsheet analysis of USAF aircraft sortie rates under a variety of assumptions about range to target, aircraft speed, and basing options. This analysis is based on historical F-15 and F-16 maintenance data and statistical analysis conducted by the Boeing Corporation on the relationship between maintenance time, sorties, and sortie duration for Boeing 737 airliners.

The use of airliner data to explore the relationship between maintenance requirements, sorties, and sortie duration may seem strange. However, using these data helps us understand and represent the influence on sortie rates of two very distinct classes of aircraft system failures: those that are a function of *cycling*—turning a component on and off—and those that are a function of aggregate time in use. Avionics components, for example, tend to produce cycle-related failures, whereas hydraulic pumps fail in relation to the hours of use that they have accumulated.

The model presented here relies on a recent study of F-15 and F-16 maintenance data[2] to help predict the relationship between sortie duration (or distance to target) and maximum sortie rates for USAF

[1]J. Lawrence Hollett, "USAF Responses to Weapons of Mass Destruction Use: Standoff Tactical Airpower Projection Option," unpublished RAND research.

[2]Craig Sherbroke, *Using Sorties vs. Flying Hours to Predict Aircraft Spares Demand,* McLean, Va.: Logistics Management Institute, April 1997.

units. The model reproduces sortie rates achieved by USAF fighters operating from Saudi bases during Operation Desert Storm, ±10 percent. While our simple aircraft-sortie-generation model does not generate absolutely precise estimates of USAF sortie-generation potential, it provides estimates that are close enough to what could be achieved to allow us to estimate the costs and benefits of various adversary base-attack strategies and USAF responses.

Our model takes as its starting point the simple observation that, at any given time, an aircraft must be either in the air or on the ground. Time spent in the air is denoted *FT* (flight time), and time spent on the ground is denoted *GT* (ground time). Therefore, sortie rate (*SR*) is a function of *FT* and *GT*:

$$SR = \frac{24 \text{ hours}}{FT + GT} \qquad \text{(B.1)}$$

Ground time can be divided into the time to accomplish routine actions required before a perfectly operating aircraft can accomplish its next mission, and those actions required to repair or replace malfunctioning systems. The former we refer to as *turnaround time* (*TAT*) and the latter as *maintenance time* (*MT*). Since $GT = TAT + MT$, we get the following equation by substitution:

$$SR = \frac{24 \text{ hours}}{FT + TAT + MT} \qquad \text{(B.2)}$$

Table B .1 summarizes the average time required for various routine TAT tasks and shows the basis for our 180-minute (3-hour) constant TAT.

Flight time is the distance from an aircraft's base to the target and back, divided by the average cruise speed:

$$FT = 2 \times \frac{\text{distance to target}}{\text{average cruise speed}} \qquad \text{(B.3)}$$

Table B.1

Average Time Required for Various Turnaround-Time Tasks

Turnaround Time (TAT) Major Actions	Average Time Required (minutes)
Land and Taxi	10
Make Aircraft Safe for Ground Ops	5
Shut Down Systems	2
Conduct Post-Flight Inspection/Debrief	15
Re-arm	50
Service	20
Refuel	30
Conduct Pre-Flight Inspection	15
Start Engine	5
Perform Final Systems Check	5
Arm	5
Taxi	10
Wait in Queue	5
Take Off	3
TOTAL	180

NOTE: These turnaround times represent typical performance of USAF maintenance personnel in force-employment exercises conducted during the late 1980s and early 1990s, as determined through interviews of senior F-15 and F-16 maintenance personnel by J. Lawrence Hollett in 1995. While it may appear that substantial time could be saved by performing the post-flight inspection, re-arming, service, and refueling operations in parallel, safety considerations prevent doing so. When refueling or re-arming operations are in progress, only fuels and munitions personnel are permitted near an aircraft.

Recent analysis of the relationship between sorties, sortie duration, and maintenance requirements for F-15 and F-16 aircraft revealed that there is a constant average of 3.4 hours of maintenance time per sortie and an additional 0.64 hour of maintenance time for every hour the aircraft is in the air.[3] This relationship yields the following formula for maintenance time:

$$MT = 3.4 \text{ hours} + 0.68FT \qquad \text{(B.4)}$$

[3]Author's conversation with Craig Sherbroke of the Logistics Management Institute, McLean, Va.

So, for a 24-hour aircraft flying a mission against a target 500 nautical miles (nmi) from its base with a cruise speed of 500 knots, the sortie rate would be

$$\begin{aligned} SR &= 24 \text{ hours}/(1.68FT + TAT + 3.4) \\ &= 24 \text{ hours}/[1,000 \text{ nmi}/(500 \text{ nmi/hour})] + 3 \text{ hours} \\ &\quad + 3.4 \text{ hours} + [0.68 \times 1,000 \text{ nmi}/(500 \text{ nmi/hour})] \\ &= 24 \text{ hours}/(2 \text{ hours} + 3 \text{ hours} + 3.4 \text{ hours} + 1.36 \text{ hours}) \\ &= 24 \text{ hours}/(9.76 \text{ hours}) \\ &= 2.45 \text{ sorties per day} \end{aligned} \tag{B.5}$$

For 12-hour, "day or night only" aircraft (such as F-117s or A-10s), sortie rates are rounded down to the integer number of sorties an aircraft could fly in a 14-hour period.[4] Using the same cruise speed and distance to target from the above example gives a day-or-night sortie rate of 2, because an aircraft flies a sortie and is ready for relaunch in 9.76 hours. Its second sortie requires 2 hours, for a total of 11.76 hours. There is not enough time to complete the 7.76-hour average ground cycle before darkness or light makes further operations impractical for the particular aircraft type, so only 2 sorties can be flown.

Note that, as the distance between an aircraft's base and its target increases, sortie rate decreases, because of two factors. The first and most obvious is that flight time has increased. However, in this model, the increase in flight time also leads to an increase in maintenance on those systems sensitive to sortie duration. So, not only do maintenance personnel have less time to work on a given aircraft as sortie duration increases (the more time it spends in the air, the less it spends on the ground), but they also have more work to do. As a result, sortie rate is *not* a linear function of distance to target.

---

[4]This assumes an average 12-hour night with 30 minutes each of twilight at dawn and dusk and 30 minutes of transit in darkness or daylight over friendly territory at both dawn and dusk.

Appendix C

FAST, LONG-RANGE-ATTACK AIRCRAFT

In Chapter Five, we recommended a long-term stand-off option involving fast, long-range aircraft. In this appendix, we present an operational concept for long-range-attack operations to stimulate thought about, and a more thorough investigation of, such options available to the United States Air Force (USAF) in the early part of the next century.

WHY FAST ATTACK AIRCRAFT?

All other things being equal, if it is less expensive to develop and deploy sophisticated weapons than it is to develop sophisticated delivery platforms, then the USAF should adopt a combination of sophisticated, long-range weapons and relatively unsophisticated delivery platforms. An example of this approach to the long-range-attack problem might be thousands of advanced cruise missiles launched from Boeing 747 arsenal planes from outside the range of enemy defenses. However, if developing and deploying sophisticated launch platforms with less-sophisticated (and less-expensive) weapons is less costly than building thousands of advanced cruise missiles, then the best solution might be something like a Mach 2 penetrating bomber armed with accurate freefall weapons such as the Joint Direct Attack Munition (JDAM).

The arsenal plane concept has been explored by others.[1] The remainder of this appendix explores some of the technical issues surrounding the Mach 2 example given above.

OPERATIONAL CONCEPT

In the following analysis, we employ a force of supersonic bombers with a maximum unrefueled range of 3,250 nautical miles (nmi) to accomplish an 8,000-nmi combat mission in the following way. First, a force of tanker aircraft escorted by F-22s and the Airborne Warning and Control System (AWACS) departs an operating base. This base would be equipped with large, submunition-resistant shelters for all aircraft and personnel and be defended by anti–ballistic missile and anti–cruise missile systems.[2] Approximately 2 hours after that departure, the bomber force launches from the same base and proceeds to a rendezvous with the refueling force 2,250 nmi from base at Mach 2, at approximately 60,000 feet.[3] At this point the bombers descend to about 40,000 feet and slow to 500 knots for 30 minutes to refuel. Bombers and tankers are protected by F-22s during this delicate operation. Following the refueling, the bombers climb and accelerate to Mach 2, fly 1,500 nmi to their targets, then egress 1,500 nmi to a second refueling rendezvous with a different group of tankers, fighters, and AWACS. The second refueling would occur approximately 5 hours into the bomber mission (bombers maintain a 30-minute fuel reserve in case this second rendezvous is delayed). After refueling, the bombers again accelerate to Mach 2 and proceed back to base. Total bomber mission time is approximately 8 hours. The F-22 mission time is also approximately 8 hours

---

[1]The best exploration has been by The Boeing Company, *747 Air-Launched Cruise Missile System Concept,* Seattle, Wash., April 1974.

[2]The structures and systems at this base could also be designed to sustain operations under chemical and/or biological attack.

[3]For a given speed and size of aircraft, it is possible to calculate an optimal altitude for maximum range. The faster the aircraft, the more drag it produces. Air density and drag decrease with altitude. Therefore, the faster an aircraft flies, the higher the optimal altitude will be. Large aircraft capable of sustained Mach 2 speeds, such as the TU144 and Concorde, cruise at 60,000 feet, whereas the Mach 3+ SR-71 cruises at 80,000 feet or higher. Since our proposed aircraft are approximately the same weight as the Concorde, we used a similar flight profile.

(due to supercruise capability); tankers and AWACS are airborne for 10 hours. Figure C.1 illustrates this operational concept.

The unrefueled range of the aircraft proposed here is much shorter than the unrefueled range of current subsonic USAF bombers. For example, the B-52H has an unrefueled range of approximately 8,500 nmi and weighs about 500,000 pounds at maximum takeoff weight. Theoretically, it might be possible to build a Mach 2 bomber with a range of 8,000 nmi; however, it would be so big that it would be impracticable. Transonic and supersonic flight regimes have enormous drag, making the thrust requirements and, in turn, the fuel requirements for supersonic aircraft much larger than those for subsonic aircraft with comparable range. Due to the "spiral of requirements" described in the next paragraph, a supersonic bomber with an 8,000-nmi unrefueled range would be so large—much heavier than a Nimitz-class aircraft carrier—that it would be ridiculous to even consider building such an aircraft.[4]

RAND *MR1028-C.1*

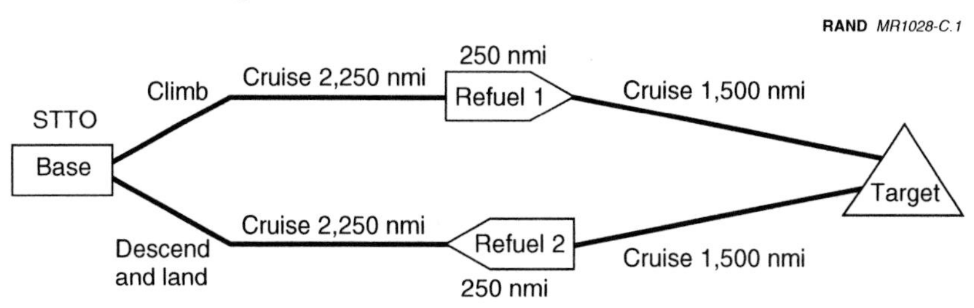

NOTE: STTO = start, taxi, take off.

Figure C.1—Operational Concept for a Long-Range Supersonic Bomber Force

[4]We used an aircraft-sizing model developed by one of our RAND colleagues, Dan Raymer, to estimate the size, weight, and amount of fuel required per mission for supersonic aircraft of various ranges and payloads. The model allowed us to input the desired range, speed, and payload, and used specific fuel consumption, aspect ratio, and lift divided by drag (L/D) to estimate the aircraft characteristics mentioned above. See Daniel P. Raymer, *Aircraft Design: A Conceptual Approach*, Washington, D.C.: American Institute of Aeronautics and Astronautics, 1989, for a complete description of equations and inputs to this aircraft-sizing model.

BOMBER CHARACTERISTICS

Figure C.2 shows the number of tankers required per bomber mission as a function of bomber range for bombers with 10,000- and 20,000-pound payloads. It is perhaps counterintuitive that bombers with longer ranges require more tankers per sortie than do those with shorter ranges. However, to achieve additional range, aircraft must carry more fuel. If we want to increase the unrefueled range of a 350,000-pound supersonic bomber with a 20,000-pound payload from 3,250 to 4,250 nmi, simply adding one-third more fuel will not do the job. Fuel is heavy—about 6.5 pounds per gallon—and takes up space within the aircraft. The additional weight and space mean a larger bomber is required to accommodate the extra fuel, and a larger wing and more power are required to compensate for the added weight and drag of the additional fuel and structure. Additional power means still more fuel is needed to achieve the increased range. This leads to yet more structure, drag, lift, power, fuel, etc. In

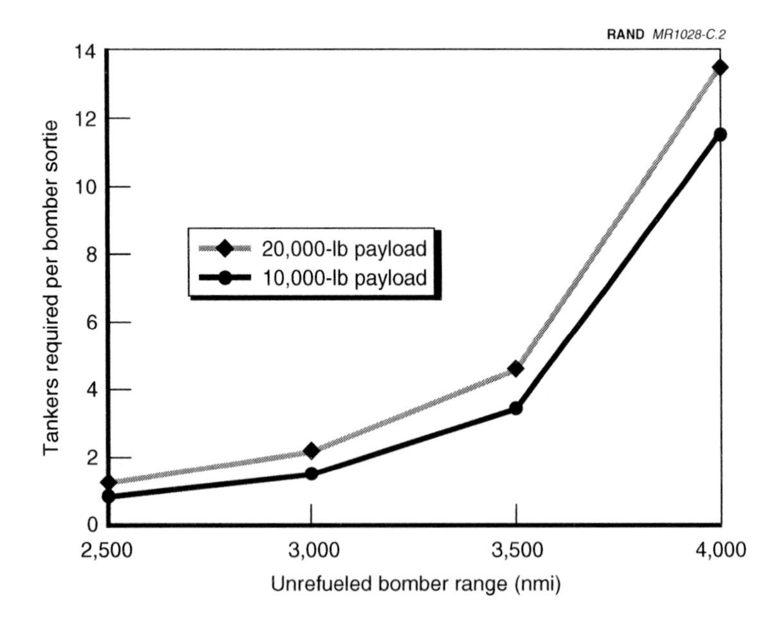

Figure C.2—Tankers Required per Bomber Sortie As a Function of Unrefueled Bomber Range

the end, this spiral of requirements leads to an increase in bomber weight from about 350,000 pounds to about 2,200,000 pounds—a 630-percent increase in weight to achieve a 33-percent increase in unrefueled range. The difference in fuel capacity is even greater. The 4,250-nmi-range bomber would have a fuel capacity of about 1,350,000 pounds, whereas the 3,250-nmi-range bomber would require a capacity of only about 197,000 pounds—a difference of 685 percent. Therefore, although the larger, longer-range bomber could fly a given mission with 33 percent fewer refuelings, each refueling would require far more fuel, necessitating more tanker support than the shorter-range bomber.[5]

Once we have established the desired range of our proposed supersonic bombers, payload drives bomber size. Obviously, the more each bomber can carry, the fewer bombers are needed to attack a given set of targets—an argument for very large payloads. However, as with long range, large payloads lead to large and expensive aircraft. In addition, there is a limit to the number of targets a crew can be reasonably expected to plan for and attack on a given mission. For example, Air Combat Command currently plans to have its B-1 and B-2 crews attack no more than 8 to 12 targets per mission.

Data from USAF precision-guided-munition (PGM) attacks during Desert Storm indicate that, when the USAF attacked a target with PGMs, it used, on average, 1.6 weapons, or a total weight of 1,700 pounds per target.[6] If we assume that the USAF uses the same average number and weight of PGMs to attack targets in future conflicts as it did on average during Desert Storm, an aircraft capable of attacking 8 to 12 targets with PGMs would require a payload of between 15,000 and 20,000 pounds. Figure C.3 shows the predicted relationship between payload and bomber weight for 3,250-nmi-range

---

[5]But only up to a certain point. As unrefueled bomber range decreases, the tankers must refuel the bombers closer to the target and farther from their base. As the tankers fly farther and farther from their base to make off-loads, more tankers are required to deliver a given amount of fuel to the bombers, since the tankers must fly farther both to and from the rendezvous.

[6]Eliot A. Cohen, ed., *Gulf War Airpower Survey* [GWAPS], Washington, D.C.: U.S. Government Printing Office, Vol. V, 1993, pp. 418, 467, 514, 606. These attacks include strikes using laser-guided bombs, Maverick missiles, and electro-optically guided bombs.

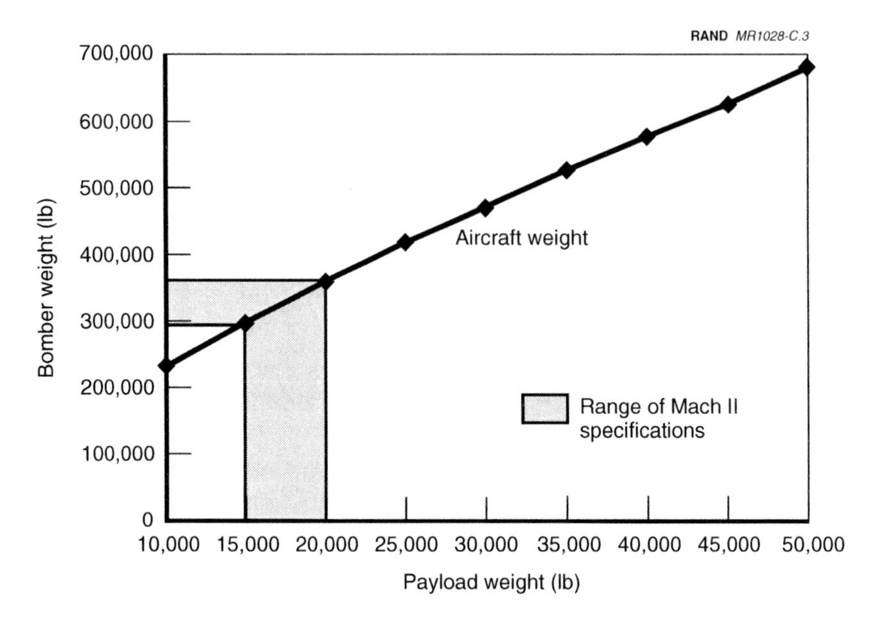

Figure C.3—Payload Versus Bomber Weight

supersonic aircraft. Aircraft capable of carrying 15,000 to 20,000 pounds of PGMs over this distance at Mach 2 would weigh between approximately 290,000 and 350,000 pounds—about 60 to 70 percent the weight of a B-2.[7]

SURVIVABILITY

We need to emphasize here that our reason for suggesting supersonic bombers has more to do with increasing the survivability of the aircraft, facilities, and support personnel on the ground than with increasing the survivability of the aircraft during the course of their

---

[7]By the time the proposed bomber reaches initial operating capability (IOC), the USAF may have fielded the family of small, smart munitions that is currently on the drawing board. If the proposed bomber used these weapons, which purport to provide the punch of a 2,000-pound bomb in a 500-pound package, it would need only a 4,000–5,000-pound payload to attack 8 to 10 targets. A supersonic bomber with 3,250-nmi range and a 5,000-pound payload would weigh only 154,000 pounds—about half as much as the aircraft under discussion here. This reduced size would lead to marked reductions in initial cost and tanker support requirements, among other things.

combat missions. This is a very different motivation from the one that led the USAF to develop supersonic bombers such as the B-58 and XB-70 in the late 1950s and early 1960s: overcoming increasingly sophisticated and capable Soviet interceptors and surface-to-air missile (SAM) systems by dramatically increasing the cruise speeds and altitudes of its bomber force. In the era before the advent of stealth technology, the Soviets showed that they could build radar-guided SAM systems capable of intercepting any high-performance bomber the United States could possibly develop. As a result, the XB-70 was canceled and the B-58 withdrawn from service.

To survive adequately against future surface-to-air and air-to-air defenses, our proposed long-range strike aircraft would have to incorporate features that minimize its radar and infrared signatures. With the F-117, B-2, and F-22 behind it, the Air Force and the U.S. aircraft industry probably possess most of the expertise needed to ensure the requisite degree of radar stealthiness. However, managing the infrared (IR) signature of an aircraft cruising at Mach 2 will present some new challenges.

The heat generated by flight at Mach 2 might make it impossible to produce a Mach 2 aircraft with a low-IR signature (the Concorde's nose cone reaches 260°F during Mach 2 cruise at 60,000 feet). It might be possible to design and build a long-range, high-altitude SAM system using an Infra-Red Search and Track System (IRSTS) to detect and track the bombers and IR guidance for the missiles. However, no such system currently exists, and the maximum detection range of such a system would be limited by atmospheric absorption of the bomber's radiated heat. If effective detection range is short, the number of SAM sites required to construct an effective barrier would be large. SAMs used as point defenses could be defeated by equipping the bombers with stand-off weapons with ranges that allowed them to attack targets from outside the range of the defenses.

HOW MANY BOMBERS?

How many supersonic bombers might it take to replicate an offensive air campaign of the size and intensity of operation Desert Storm? During the course of the 43-day conflict, USAF aircraft conducted 28,295 strikes—an average of 660 strikes per day. Only about 7,800 of these attacks—about 28 percent—were made with PGMs.[8] However, to maximize the impact of our bomber force, we propose to attack the same number of targets using only PGMs. As mentioned in the preceding section, the USAF used 1,700 pounds of PGMs in each PGM strike. To attack 660 targets per day with 1,700 pounds of PGMs each, our supersonic bomber force would need to deliver about 560 tons of PGMs per day. With each aircraft carrying between 7.5 and 10 tons of PGMs, an effort of this size would require 56 to 75 bomber sorties per day.[9] Figure C.4 summarizes this information. Maintaining a ready strike force of 56 to 75 bombers while allowing for training aircraft, test aircraft, aircraft in depot maintenance, etc., would probably require a total force of 80 to 105 aircraft.

SUMMARY

For the supersonic bomber concept described here, a total inventory of approximately 80 to 105 Mach 2 bombers with the following characteristics could deliver enough PGMs (about 560 tons per day) to replicate the USAF Desert Storm effort:

- an unrefueled range of 3,250 nmi
- a weight of 290,000 to 350,000 pounds each
- a payload of 15,000 to 20,000 pounds
- support of 37 to 40 percent of the current USAF tanker fleet and 100 air superiority fighters.

In addition, Mach 2 bombers could attack targets almost anywhere in the world while operating from well-protected, permanent bases on U.S. and UK territory.

---

[8]Cohen, GWAPS, 1993, Vol. V, p. 418. We use the same definition of a *strike* as presented in Cohen, GWAPS, 1993, Vol. V, p. 403: an attack by a single aircraft on a single target. The aircraft may use one or more weapons in its attack and, if it has enough weapons, may conduct multiple strikes in a single sortie.

[9]On the basis of KC-135R and KC-10 mission profiles computed from their respective flight manuals and assuming an 80-percent tanker mission-capable rate, we estimate that sustaining this level of bomber activity would require the commitment of between 37 and 40 percent of the USAF's total existing tanker capacity. Additional tankers would be required to support escorting fighters, AWACS, etc.

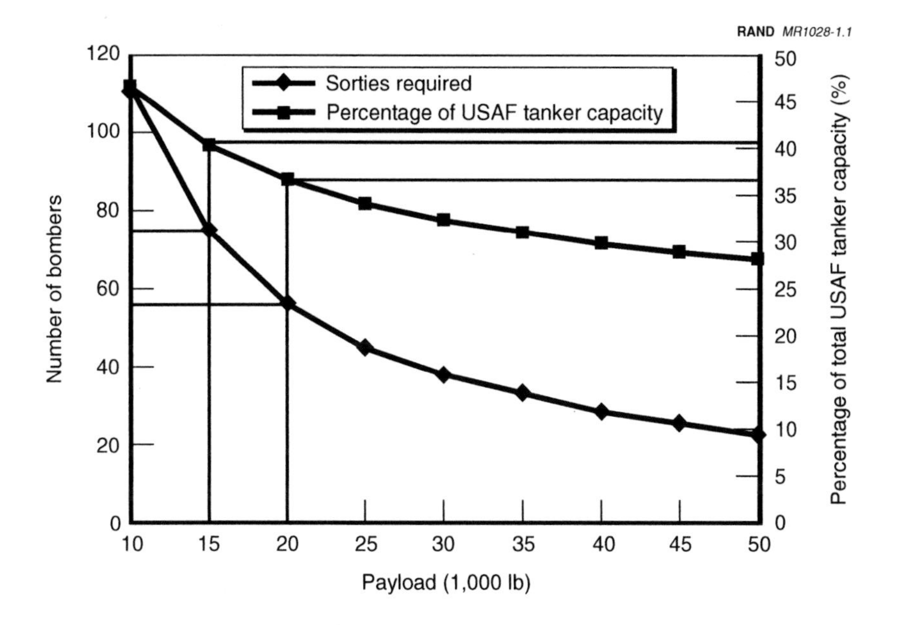

Figure C.4—Bomber Sorties and Tanker Capacity Required As a Function of Bomber Payload

BIBLIOGRAPHY

The Boeing Company, *747 Air-Launched Cruise Missile System Concept,* Seattle, Wash., April 1974.

Cohen, Eliot A., ed., *Gulf War Air Power Survey, Vol. V: A Statistical Compendium and Chronology,* Washington, D.C.: U.S. Government Printing Office, 1993.

Chow, Brian, *Air Force Operations in a Chemical and Biological Environment,* Santa Monica, Calif.: RAND, DB-189/1-AF, 1998.

Douhet, Giulio, *The Command of the Air,* Washington, D.C.: Office of Air Force History, 1983 (originally published in 1921).

Frost, Gerald, *Operational Issues for GPS-Aided Precision Guided Weapons,* Santa Monica, Calif.: RAND, MR-242-AF, 1994.

Frost, Gerald, and Irving Lachow, *Satellite Navigation-Aiding for Ballistic and Cruise Missiles,* Santa Monica, Calif.: RAND, RP-543, 1996a.

———, *GPS-Aided Guidance for Ballistic Missile Applications: An Assessment,* Santa Monica, Calif.: RAND, RP 474-1, 1996b.

Hart, Basil Henry Liddell, *Thoughts on War,* London, 1944.

Hogg, Ian V., ed., *Jane's Security and CO-IN Equipment,* Coulsdon, Surrey, England: Jane's Information Group, 1991–1992.

Jackson, Paul, *Jane's All the World's Aircraft,* Coulsdon, Surrey, England: Jane's Information Group Limited, 1997–1998.

Joint Chiefs of Staff, *Department of Defense Dictionary of Military Terms,* Washington, D.C., Joint-Pub 1-02, March 23, 1994.

Lennox, Duncan, ed., *Jane's Strategic Weapon Systems*, Coulsdon, Surrey, England: Jane's Information Group, Issue 24, May 1997.

Lennox, Duncan S., and Arthur Rees, eds., *Jane's Air Launched Weapons*, Coulsdon, Surrey, England: Jane's Information Group, Issue 12, 1990.

Munson, Kenneth, ed., *Jane's Unmanned Aerial Vehicles and Targets*, Coulsdon, Surrey, England: Jane's Information Group, 1995, 1996, 1997.

National Imagery and Mapping Agency, *DOD Flight Information Publication: High and Low Altitude Europe, North Africa, and Middle East*, St. Louis, Missouri, February 27, 1997.

The Nonproliferation Review, Spring–Summer 1995, Vol. 2, No. 3, pp. 203–206.

Prangue, Gordon W., *At Dawn We Slept*, Norwalk, Conn.: The Easton Press, 1988.

Raymer, Daniel P., *Aircraft Design: A Conceptual Approach*, Washington, D.C.: American Institute of Aeronautics and Astronautics, 1989.

Sherbroke, Craig, *Using Sorties vs. Flying Hours to Predict Aircraft Spares Demand*, McLean, Va.: Logistics Management Institute, April 1997.

Shlapak, David, and Alan Vick, *"Check Six begins on the ground": Responding to the Evolving Ground Threat to U.S. Air Force Bases*, Santa Monica, Calif.: RAND, MR-606-AF, 1995.

U.S. Air Force, *Facility Requirements*, Air Force Handbook AFH 32-1084, September 1, 1996.

US Air Force Statistical Digest, 1996.

Vick, Alan, *Snakes in the Eagle's Nest: A History of Ground Attacks on Air Bases*, Santa Monica, Calif.: RAND, MR-553-AF, 1995.